THE TRUTH ABOUT
RAINFOREST DESTRUCTION

THE TRUTH ABOUT RAINFOREST DESTRUCTION

Russell G. Coffee

Better Planet Press • Austin, Texas

Copyright ©1996 Russell G. Coffee
ALL RIGHTS RESERVED

First Printing

Published by The Better Planet Press, P.O. Box 160146, Austin, Texas 78716

Library of Congress Catalog Card Number 94-71634
International Standard Book Number 0-9641721-0-0

Printed on Recycled Paper
Printed in the United States of America

Cover photos taken by space shuttle astronauts and reprinted with permission of NASA. Front-the smoke plumes of slash and burn farmers in Indonesian Borneo. Back-smoke plumes rising over Mozambique's Zambezi River.

To order additional copies see page 115

There is in us all the dark conscience of our murder of the primeval forests—of something in their depth which is a depth in us; of our refusal of their value, of our disdain of the red man who was the spirit of those forests and who is yet, beneath the layers of law and memory, the spirit of ourselves.

—*Waldo Frank (1889-1967)*
American Novelist/Social Critic

Many thanks to Alice, Monica, Kris, Scott, and Catherine for their support.

Contents

INTRODUCTION	**9**
PREFACE	**12**
THE MEANING OF EXTINCTION	**17**
HOW BAD IS IT?	**23**
DEFORESTATION RATES	23
SPECIES EXTINCTIONS	24
LOST PHARMACIES	25
THE HUMAN TOLL	26
CLIMATE CHANGE	27
SURPRISES	28
CAUSES OF RAINFOREST DESTRUCTION	**30**
CATTLE RANCHING	30
LOGGING	32
SMALL-SCALE FARMING	34
A PROCESS OF DEFORESTATION	37
THE UNDERLYING CAUSE	**39**
ORIGINS OF A DEBT CRISIS	40
THE DEBT CRISIS BEGINS TO FUEL DEFORESTATION	45
WHAT IS THE SITUATION NOW?	49
COLONIALISM?	51
THE WORLD ON FIRE	**53**
A CONFLICT OF CULTURES	**69**
THE PENAN BATTLE INTERNATIONAL LOGGING COMPANIES	72
THE YANOMAMI AND THE GOLD RUSH OF THE 1980s	75
ENVIRONMENTAL ORGANIZATIONS ON THE WAR-PATH	78
THE KAYAPO STOP THE WORLD'S LARGEST DAM PROJECT	79
THE KUNA SAVE THEIR RESERVE FROM SLASH AND BURN	83
NEW DIRECTIONS	**87**
RAINFOREST PRESERVES THAT PROFIT	89
SUPPORTING SUSTAINABLE BUSINESSES	91
GOVERNMENT REGULATIONS	94

Rethinking Intergovernmental Lending	100
Resolving the Debt Crisis	103
The Sustainable Development Solution	107
LEADERS IN THE COMING ECOLOGICAL ERA	**108**
Political Leadership	108
The Role of the Non-Governmental Organization	109
Leadership in the Business World	110
Grass-Roots Support	110
Individuals as Leaders	112
TO ORDER ADDITIONAL COPIES	**115**
APPENDIX	**116**
Organizations Working to Preserve Rainforests	116
REFERENCES	**120**

INTRODUCTION

This is a book meant to spread awareness about a painful reality—the destruction of rainforests is wiping out plant and animal species at a rate unknown since the passing of the dinosaurs.

When I first heard about this, I could not believe it. As a child, I had always assumed that somebody (possibly the president) was in charge of the world's environment. It never occurred to me that a mass extinction could happen again.

It is happening again. Our rainforests are shrinking, our climate is changing, and this year alone nearly 30,000 species will disappear forever.

The Truth About Rainforest Destruction is not about statistics, however. It is more the story of our culture's attempt to prosper at the expense of the natural world.

As you read this story, you will discover why rainforests are so important, why they are being destroyed, and who is benefiting from this destruction.

You also will learn how tropical deforestation is affecting several Indian tribes that live in the rainforest—tribes which are today fighting the last of the Indian wars.

Finally, you will read about solutions—real solutions. Conventional wisdom holds that if we recycle enough products or simply conserve and consume less, then our environmental problems will be solved. This is false. Never in history has mankind reduced its consumption of resources. It is doubtful we will start now.

To make real environmental progress, we must make changes in our economic system. These economic changes—a new environmentalism—I will explain at length while introducing you to the struggles of men and women determined to stop deforestation.

Dedicated to Chico Mendes

PREFACE

WHAT WE KNOW

Most of us experience our natural surroundings as being relatively stable. The few changes in the environment that we observe in our daily lives seem of little significance.

In the context of evolutionary history, however, what we see in a lifetime is the geologic blink of an eye—and may be more threatening than we realize.

Consider our world's condition in the last measure of geologic time, called the Mesozoic era, and the radical changes that occurred when this era came to an end.

The earth in this earlier time had different continents, different animals, and another ecosystem. South America was a part of Africa. Plants did not produce flowers. Dinosaurs roamed the earth.

Among the dinosaurs was the flying Pteranodon, the largest creature ever to fly, and the Brachiosaur—as tall as a four-story building—the tallest animal ever to walk. One hundred million years ago, these beings and others like them lived all over our planet.

Around 65 million B.C. something changed. These exotic life forms disappeared, the Mesozoic era ended, and about half the earth's species became extinct. In a twist of evolutionary fate, the dinosaurs all died.

Why dinosaurs died is unknown. What caused their environment to deteriorate, we can only speculate.

Some theorists contend the earth in this former time was bombarded with thousands of meteors, some having an impact roughly equivalent to one of today's nuclear

bombs. Others say an enormous volcano erupted and emitted enough soot to block out sunlight for several years. Possibly a comet narrowly missed the earth, temporarily reversing the planet's magnetic field.
  Still others argue that maybe one species adapted all too well. Initially prospering in a rich environment, this species may have hunted its food sources to extinction, thus destroying itself, the food chain, and the balance of nature.
  We may never know which theory is correct. The extermination of millions of species in the span of a thousand years—possibly only a few years—is a mystery that has never been solved. Most likely, the creatures that perished in this time did not even know why they were dying.

  Today, we live in a different world, with different continents, other species, and another environment. From the perspective of a single lifetime, this made-anew world appears stable. But closer examination reveals something else—today's Cenozoic environment also is dying.
  Most of the world's water supply is being polluted. Our protective ozone layer is being depleted. The air in our major cities is becoming less breathable. Our jungles are becoming deserts.
  Our planet is entering another period of mass extinction. In Western Germany, for example, 34 percent of all invertebrate species are on the verge of disappearing. In England, 17 percent of insects are classified as threatened or endangered.
  In the oceans, all species of whales may soon disappear, as may all eight species of sea turtles.

In the Western Hemisphere, 33 percent of parrots are threatened with extinction.

Ducks also are vanishing. Pintails have declined by 80 percent in the last 40 years; Black Ducks are down by 62 percent.

In Africa, there were 65,000 Black Rhinos in 1970; now, they number only 2,300, a decrease of 96 percent.

At the turn of the century, there were more than 100,000 tigers in existence; today, there are fewer than 6,000.

In the 1980s, half of all the African elephants perished. In that decade, one elephant died every 8 minutes.

Worldwide, 25 percent of all bird species can no longer be found. Also worldwide, about 20 percent of freshwater fish are extinct.

Making matters worse, these rates of extinction are increasing. Since the beginning of human civilization, about 20 percent of all species, both plant and animal, have disappeared; within the next few decades, this figure is expected to reach 40 percent. Today, six species are being extinguished every hour.

Unlike the mass extinction of the dinosaurs, in this modern-day extinction there is a difference. This time, a species exists that can pass knowledge from one generation to another, that can communicate worldwide, that has an information system of computers, satellites, TVs, radios, newspapers, and books. This time, a species exists that knows why.

In today's era of extinction, Homo sapiens know why things happen, as they happen. For instance, we know that many of the products we use release chemicals that damage the ozone layer. We know that a thinning ozone layer is contributing to a worldwide population decrease of amphibians.

We know that Taiwanese fishermen using 40-mile-long driftnets are sweeping the oceans of fish and other life. The United Nations reported that these nets drowned between 300,000 and 1 million dolphins worldwide in the 1989-90 fishing season alone.

We know that Eastern Europe has sub-standard nuclear power plants that leak radiation. We know that for years the Soviets have been dumping spent nuclear reactors at sea.

We know that right now fires are raging in tropical forests. A single satellite photo shows 6,000 such fires burning simultaneously in the Amazon region alone. We know that these fires are deliberately set to make way for cattle and crops—products that we, in more developed countries, will later consume. We know that these fires are the greatest cause of our modern era of extinction.

Because of what we know, we can foresee our own demise. Our governments keep statistics on the toxicity of rivers. Our computer scientists create models of ozone holes and estimate climate changes as these holes expand.

The United Nations Food and Agriculture Organization charts the world's rate of deforestation—a rate that shows an area of rainforest the size of a football field being destroyed every second. We know how many football fields of rainforest are left.

We know that cleared rainforest lands are being damaged; overgrazing and erosion are reducing their ability to regrow vegetation. Through aerial photography we are watching the encroachment of deserts into these once-fertile lands. We know how much fertile land is left.

These things we know; they are facts. They threaten our existence and the existence of every living thing on the planet.

Today's era of extinction is indeed unique. For unlike the dinosaurs that perished in ignorance, we the human species know why our environment is dying. Our technology has collected this information, our scientists have analyzed it, and our global communication systems have brought it into our homes.

THE MEANING OF EXTINCTION

I will argue that every scrap of biological diversity is priceless, to be learned and cherished, and never to be surrendered without a struggle.

—Edward O. Wilson

I don't remember how old I was when I first learned about dinosaurs, but I do remember studying them one day in the second grade. They were fascinating to me: huge, seemingly ferocious monsters, living in swamps.

I left school that day with some apprehension. If these creatures had existed at one time, where were they now? I thought about the prospects of meeting one as I walked home. Was there a Tyrannosaurs rex in the wooded lot where we played cowboys and Indians? Perhaps there was a Stegosaurus under the Shoal Creek bridge?

It wasn't until I was older that I was able to fully understand the meaning of extinction: when a species becomes extinct, it no longer exists; it will not be born again; it cannot even evolve again.

Realizing this, I was relieved but also sad. I knew I could walk the streets without fear of a dinosaur pouncing on me, but I also knew, that no matter where or how far I walked, the only dinosaurs I would find would be living in my school books.

At the age of 15, I saw my first rainforest, and my fascination with mysterious creatures returned. Some parents and teachers from my high school had organized a trip to the south of Mexico, and I was one of a dozen students fortunate enough to go.

Shortly after we arrived, our group boarded a bus and headed into the jungle interior. After about two hours of driving over potholes and winding down dirt roads, we stepped off the bus and into the shadows of a rainforest.

As we walked among the vegetation, our tour guide told us about the various types of rainforests and the exotic creatures, plants, and animals they contain.

Along the equator are equatorial rainforests. These forests are always warm—about 80 degrees Fahrenheit—and always damp—raining almost every day. In some equatorial rainforests it rains more than 300 inches a year.

We were in a subtropical rainforest. This type of rainforest is located some distance from the equator but still within the tropics. Subtropical rainforests have two seasons, wet and dry.

Also in the tropics are montane and mangrove rainforests. Mangrove forests are more like dense thickets along muddy coastal areas. Montane forests, often called cloud forests, are found at altitudes above 5,000 feet and are frequently shrouded in mist.

Outside the tropics are mostly temperate rainforests, which have a seasonal climate. These rainforests are located, among other places, in New Zealand, Japan, South Africa, and along the Pacific coast of the northwestern United States and Canada.

As we walked deeper into the jungle I looked high overhead. Never had I seen such tall trees!

THE MEANING OF EXTINCTION 19

The tallest trees in a rainforest are called emergent trees. In equatorial rainforests, these trees may tower 160 feet or more.

The tops of emergent trees are considered to be one of four distinct layers of rainforest vegetation. These tree tops jut randomly above the next layer of vegetation, called the canopy.

The rainforest canopy is a continuous cover of branches and leaves about 100 to 130 feet above the ground. Almost two-thirds of all plants and animals in a rainforest live within this strata.

Below the emergent trees and the canopy is the understory. This layer consists of smaller trees that grow as high as 50 to 80 feet above the ground.

Below all of this is the forest floor, an area completely shaded by the upper layers and having little vegetation. Only a few seedlings, small plants, and decaying plant matter can be found here; rarely is there grass or any ground cover.

Almost all the trees in a rainforest are connected by an endless tangle of vines and a myriad of plant life. Most of this plant life lives entirely within the canopy, never touching the ground. Much of it has never been studied.

Rainforest explorer William Beebe remarked in 1917, "Yet another continent of life remains to be discovered, not upon the earth, but one to two hundred feet above it."

After walking for 30 minutes, our group reached its destination: a high stone wall overgrown with vines. As we approached, lizards and insects scurried into its crevices.

On the other side of the wall were the crumbling ruins of an ancient Mayan city. For about an hour we walked among these ruins as our guide told us about the Mayan culture.

The Mayans were one of the great ancient civilizations of the New World. Their culture flourished within the rainforests of Mexico and Central America for almost 1,500 years.

The Mayans designed and built ornately carved temples and created a written language. They also devised a 365-day calendar and the mathematical concept of zero. At its apex, Mayan civilization supported some two million people.

Around 900 AD the Mayans mysteriously disappeared. No more temples were built. No more Mayan books were written. Their cities were abandoned. Why the Mayans vanished is one of the great mysteries of anthropology.

Equally mysterious is how the Mayans managed to support themselves for 1,500 years before they disappeared. To do this, they must have had an elaborate system of rainforest agriculture. But, rainforest soils are too poor to support extensive crop or livestock production; even with modern technology, our culture is unable to farm or ranch the region for very long. What did the Mayan farmer know about using and maintaining rainforests that has eluded us?

The answer may lie with today's rainforest Indians, who, like the Mayans, use the region without destroying it. Understanding how these Indians sustain themselves may be the key to preserving rainforests. Tribes of these Indians are vanishing rapidly, however.

On our way back to the bus, our guide asked us to notice the many colors and sounds of the jungle. In the distance, we could hear the screeching calls of Howler

THE MEANING OF EXTINCTION

monkeys while overhead fluttered brightly colored macaws and a green Quetzal.

The Quetzal, the national bird of Guatemala, was once worshiped by the Aztecs. (Only Aztec royalty could wear the Quetzal's long, iridescent green tailfeathers, and these feathers had to be plucked without injuring the bird.)

Any one tree might hold as many as 40 different species of ants, and the entire jungle contained over 200 species of poisonous snakes. Somewhere in the brush were many of the world's last wild cats—the jaguar, ocelot, puma, jaguarundi, and the Margay.

Our guide also told us about rainforest animals in other parts of the world: the rarely seen, pink, freshwater dolphins that live in the Amazon River; the Queen Alexandra butterfly of New Guinea with an 11-inch wingspan; a miniature rhinoceros called the Javen Rhino.

Our guide said that in Columbia rainforest Indians make poison blow-darts from magenta-blue frogs, whose skin secretes one of the deadliest poisons on earth; one gram is potent enough to kill 5,000 people.

In Asian jungles, he said, there are snakes that can glide from tree to tree by flattening their stomachs, making turns in mid-air by twisting their bodies.

I was filled with a sense of wonder. Here in the world's rainforests live the bizarre and mysterious creatures of another world that as a child I imagined dinosaurs to be.

Here live 30-foot anacondas and pythons that can swallow a man whole; Komodo dragon-lizards, which can grow to lengths of 10 feet and weigh 350 pounds; rodents, called capybarases, that weigh 150 pounds; giant anteaters; and the Tapir, a horse-like creature with an elephant-like snout.

Millions of exotic, mysterious creatures live in rainforests. Most have yet to be discovered.

Now, 16 years after my first visit to the rainforest, I am in another tropical country, walking through another rainforest. In this rainforest, however, there is no color and there are no sounds. The ground is barren. The sky is thick with smoke, and everything is black.

The only tree I see is a skeleton of charred branches....Ranchers are clearing new pastures.

I think about the creatures that must have lived here. Have they also gone the way of the dinosaur? Will our children have the opportunity to see rainforest creatures, or will they learn that rainforest animals exist only in school books?

HOW BAD IS IT?

The world is in great danger. When the trees die, the Earth dies. We will be orphans without a home, lost in the chaos of the storm.

— *Tacuma of the Kayapo 1988*

Most people do not give much thought to rainforest destruction. Those who do, think of it as just one of many environmental problems, which will not affect their lives.

Rainforests are being destroyed, however, and the consequences may prove disastrous for us all.

Deforestation Rates

Historically, the world's rainforests covered about 14 percent of the earth's land surface. They now cover only about six percent. This six percent is an area roughly equivalent in size to the United States. Each year, however, an area about the size of Georgia is destroyed.

This means that of the 3.04 million square miles of tropical rainforest that remain, 1.8 percent is being lost annually.

This rate of deforestation is increasing. According to the United Nations, in the late 1970s about 28,000 square miles of rainforest were destroyed each year. By the late 1980s the number had increased to almost 59,000 square miles per year (about one acre per second).

If the trend continues, the world's rainforests will disappear in all but the most remote areas within the next 25 years.

Norman Myers, an environmental consultant to the United Nations and the World Bank is a rainforest expert. He writes in his book, *Future Worlds*, "Our atlases traditionally feature a vivid green band around the equator denoting the most lush vegetation on earth. We may soon have to recolor the band a dirty brown to depict the death of the greatest celebration of nature ever to grace the face of the planet."

Species Extinctions

The destruction of rainforests means not only losing trees but also destroying ecosystems upon which trillions of living things depend.

Of the 830 billion tons of living matter on our planet's surface, 460 billion tons (55 percent) are in tropical forests. Of the 30 million species estimated to inhabit the earth, 70 percent live in the world's rainforests.

The destruction of these forests is creating the largest and fastest mass extinction of all time. Species now are becoming extinct at a rate of almost 150 a day—as much as 10,000 times the normal evolutionary rate.

This rate of extinction is increasing, as well. It is increasing not only because the rate of habitat destruction is increasing but also from a condition called "species crash."

Species crash is the death of an entire community of plants and animals caused by the death of a few key species upon which the others depend. Remove the trees from a rainforest, for example, and the many thousands of species that depend on those trees for food, shelter, and shade also will perish.

Because of species crash, in a few decades we may be witnessing the extinction of thousands of species a day and be powerless to stop it.

Unless we effect a change now, half the living things on the planet will die in our lifetime.

This is geologic history in the making. According to Myers, "This mass extinction will far surpass the 'great dying' of the dinosaurs and associated species 65 million years ago. This time, many more species will be involved, the time span will be far shorter, and there will be much more of an impoverishing impact on the future of evolution."

Lost Pharmacies

Destroying rainforests means destroying the world's largest and most diverse reservoir of plants and animals. From these living things come organic compounds of immense medicinal value, including nearly one-fourth of all prescription drugs.

Rainforest plants with medicinal properties include the Rosy Periwinkle, which produces two powerful anti-cancer agents, vinblastine and vincristine. Also, the same plant compound used by some rainforest Indians to make poisoned-tipped arrows, D-Tubocurarine, is now used as a sedative in modern operating rooms. Some rainforest plants, such as the Moreton Bay chestnut tree of Australia and the sub-tropical plant Psoralea Corylifolia, are even showing potential in the treatment of AIDS.

much of the United States, could become deserts. Most of the world's food supply could be wiped out.

A warmer climate also will cause sea levels to rise. This will occur as polar ice caps melt and the upper layers of the oceans become warmer and expand. As little as a three-foot rise in sea levels will be devastating. Coastlines would change. Cities at sea level, such as Houston, Miami, and Hong Kong, would be inundated as would many of the world's island nations.

Rising sea levels will force millions of people to migrate inland. Most coastal plants and animals, however, will drown. More extinctions will result.

Surprises

The destruction of the world's rainforests may surprise us with consequences we cannot predict. For example, trapped in the earth's frozen tundra are 10 trillion tons of methane, the second leading contributor to the greenhouse effect. Should global warming result in a melting of tundra, then this methane could be released. So much methane in the atmosphere could be enough to superheat the planet, perhaps making it uninhabitable.

Scientists are only now considering this possibility. But, it is certain that a changing climate will have unexpected consequences.

None of these cataclysmic events need occur. What if we put an end to deforestation and began reforestation? These trends toward cataclysm could reverse. Changes in climate and the next mass extinction could be averted.

This is the challenge facing our generation. We need to search for a better understanding of our environmental problems and then organize our world so that both we and the environment flourish.

CAUSES OF RAINFOREST DESTRUCTION

I am trying to save the knowledge that the forests and this planet are alive, to give it back to you who have lost the understanding.

—Chief Paulinho Paiakan of the Kayapo

Rainforests are being destroyed by mainly three things: cattle ranching, commercial logging, and the farming practices of people living in rainforests.

Of course, most ranchers, loggers, and farmers are environmentally sensitive, and our society needs the food and materials which they provide. But, when loggers relentlessly over-harvest rainforests and farmers and ranchers over-work and over-graze cleared rainforest lands, the environment is damaged because resources are consumed and not replaced.

This consumption of resources without replacement is said to be a "non-sustainable use of resources." Specifically, it is non-sustainable ranching, logging, and farming that destroys rainforests.

Cattle Ranching

Clearing rainforests to create new pastures for cattle destroys approximately 5,700 square miles of rainforest each year. This comprises about 11 percent of all rainforests destroyed annually.

CAUSES OF RAINFOREST DESTRUCTION

In some parts of the world, cattle ranching is the primary cause of deforestation. This is the case in Central America and some parts of South America. In Latin America, over the last twenty years, more than 78,000 square miles of rainforest were converted to cattle pasture, an area about the size of Nebraska.

Soil Quality
Rainforest soils are typically thin, poor, and inadequate for cattle ranching. Most nutrients are stored in the forest's vegetation. When this vegetation is removed, the soil has little regenerative power and cannot grow enough grass to sustain cattle for long.

Cleared rainforest lands do grow ample grass initially, however. This is because most ranchers clear rainforest by burning, which enriches the soil with ash. Nevertheless, soil fertility declines with each passing year.

Most ranchers of rainforests are aware of the region's poor soil quality. These ranchers, however, are content to graze cleared rainforest lands for a few years and then to abandon these pastures and clear new ones.

For this reason, ranching in rainforests has become a perpetual cause of deforestation and is considered non-sustainable. In Amazonia, one-third of the rainforest land cleared for cattle today lies abandoned.

Governmental Incentives
Because of poor soil quality and the expense of acquiring and clearing new lands, many rainforest cattle operations are economic failures. In countries like Brazil, however, governmental incentives help rainforest cattle ranchers earn profits. These incentives include subsidies

and tax breaks; some tropical governments even give away rainforest lands to those who will clear and ranch it.

Governments outside the tropics also encourage cattle ranching in rainforests. They do this by sanctioning international loans that develop the industry. In the 1970s the World Bank and the Inter-American Development Bank made over $4 billion in loans and grants to support cattle ranching in rainforests.

Because of governmental incentives, beef industries in tropical countries are booming. In the last 25 years, beef production has tripled in Costa Rica, El Salvador, Guatemala, and Nicaragua.

Ironically, these same Central American countries now consume less beef than they did a quarter century ago; their citizens cannot afford the prices that developed countries are willing to pay. In the last 25 years, most of the beef produced in Central America was exported to the United States.

Logging

Another principal cause of tropical deforestation is commercial logging. Each year, this industry destroys about 11,400 square miles of rainforest—about 21 percent of the total.

Commercial logging is primarily a problem in Southeast Asia. About 70 percent of all wood exported from the tropics comes from this region; the world's largest exporter is Indonesia.

The leading importer of tropical wood is Japan. This small island nation imports about half the wood harvested from the tropics.

The result of Japan's appetite for wood can be seen in the landscapes of its principal trading partners, which were at one time almost completely covered with forests. These partners include the Philippines, which is now 80 percent deforested; Thailand (85 percent deforested); Malaysia (50 percent deforested); and more recently Indonesia (30 percent deforested).

The Low Cost of Tropical Wood
Tropical woods, long prized for their strength and beauty, are today inexpensive to harvest.

This is because most tropical countries are underdeveloped. Compared to the costs of land and labor in industrialized countries, it costs little to clear-cut timber in tropical rainforests. Moreover, many tropical countries impose few environmental constraints on logging companies.

To governments of tropical countries, the export of wood is an $8 billion industry. Products include logs, lumber, plywood, and wood pulp. End products include many disposable items, such as paper, packaging materials, paper cups, and paper plates. In Japan tropical wood is even used to make disposable chopsticks.

Local Government Policies
Many tropical governments encourage commercial logging of their rainforests. Some Brazilian lumber companies have operated tax-free for as long as 15 years. Other Brazilian lumber companies have received business loans at very favorable interest rates.

In Malaysia the government has consistently undervalued its currency to make labor and materials less expensive to foreign firms that export wood.

The most destructive governmental policy fostering deforestation, however, is the practice of granting short-term logging leases. A company with a short-term logging lease simply harvests all it can before the lease expires and has little incentive to plant trees for future harvests.

Host countries generally make these short-term leases because the revenues they get are based on the amount of wood harvested. These countries want large, quick harvests.

When Will Non-Sustainable Logging End?

These logging practices are, of course, non-sustainable and destructive to the industry in the long run. Many countries, which today are the leading producers of tropical timber, will soon exhaust their supplies.

Largely stripped of their forests, Thailand and the Philippines already import wood. So does Laos. China and Malaysia should be logged out within the next few years. In fact, the World Bank estimates that only 10 of the 33 countries currently exporting tropical wood will have wood to export beyond the year 2000.

SMALL-SCALE FARMING

The main cause of rainforest destruction is the conversion of rainforests to small-scale farms. These small-scale farms are usually started by individuals, families or extended families to grow their own food.

Small-scale farming accounts for about 61 percent of all rainforest destruction—almost 33,000 square miles annually. This equates to an area slightly larger than the entire state of South Carolina that is destroyed each year by small-scale farmers.

CAUSES OF RAINFOREST DESTRUCTION

Slash and Burn

Most small-scale farmers who clear rainforests do so by cutting down trees and burning vegetation, a method called "slash and burn." These farmers typically move their farms to new areas in the rainforest every few years because the region's heavy rains erode cleared fields and wash nutrients from the soil. This process of clearing, farming, and moving on is called shifting cultivation. Done properly, shifting cultivation does not irreversibly damage rainforests. Indians in the rainforest sustainably farmed in this way for centuries.

These Indians, however, were few in number and they typically abandoned their farms after a single season before their fields became eroded. As a result, rainforest vegetation was able to regenerate. It is speculated that the entire Amazon jungle at one time or another was farmed by Indians.

In the mid-1960s, however, a new type of farmer began to cultivate the rainforest. These new farmers were immigrants from other parts of their countries and had no experience farming rainforests. Generally, they tried to create permanent farms, which, after a number of years, became eroded, unable to grow vegetation, and had to be abandoned.

Today, farmers are immigrating into rainforests and attempting to create permanent farms in record numbers.

Government-sponsored Immigration

Most of these immigrant farmers have no choice but to move into sparsely populated rainforests. Over-populated cities and poverty have left them no other means to support themselves.

Governments even encourage these migrations. They do this by building roads into rainforests, subsidizing businesses there, and making land grants to people willing to settle these frontier areas—not unlike the land grants that settled the western United States in the 19th century.

Transportation Infrastructure

As road networks in the rainforest expand, deforestation increases. Tropical governments build these roads to help small-scale farmers and and other forest workers immigrate into otherwise impenetrable jungles.

Among the larger highways credited with causing widespread deforestation are the 1,400-mile Transamazon Highway in northern Brazil and the 1,000-mile highway BR 364 from southern Brazil into the Amazon.

In Peru, the Carretera Marginal highway has opened up an estimated 12.5 million acres of rainforest to farming. In Africa, the 600-mile Transgabonnais railway is being built to help logging and mining companies bring workers in and move rainforest products out. Currently under consideration is the completion of the Pan American highway through the Darien rainforest near the southern border of Panama.

Free Land

Some tropical governments allow small-scale farmers to obtain land in the rainforest at virtually no cost. In the 1980s, under the Polonoroeste Regional Development Program in Brazil, settlers paid only a small title fee for land in the rainforest. These settlers could then recoup this fee and their moving expenses by selling timber from their new homesteads.

CAUSES OF RAINFOREST DESTRUCTION

In some underdeveloped countries, title to remote land in the rainforest can be established only after the land is put into productive use. In such a case, the more land a settler clears, the larger his land grant.

Low Interest Loans

Government subsidies encourage small-scale farming in the rainforest. This is the case in Brazil where settlers with title to land are eligible for low-interest government loans.

The Brazilian Central Bank's National Rural Credit System makes over 100 different types of loans to rainforest settlers, most to support livestock and agricultural operations.

Among these loans are lines of credit for buying animals and equipment, creating pastures and irrigation systems, acquiring fertilizer and pesticides, and the like. This government agency also guarantees a minimum price for certain crops grown in the rainforest.

In one of Brazil's rainforest colonization projects (the Gi-Parana project) settlers were surveyed about how they spent their loans—nearly one-fourth used these loans, in part, to buy chainsaws.

A Process of Deforestation

Often, small-scale farming, cattle ranching, and logging only partially destroy rainforests. In some instances, these partially damaged rainforests can regenerate if left fallow for a number of years. However, where these three activities happen in succession, the cumulative effect is devastating to the environment—and irreversible.

The process generally starts with loggers, who cut roads into virgin forests and harvest the biggest and best trees. Then, with easy access along these new roads, small-scale farmers (squatters) slash and burn the remaining forest and establish temporary farms. After a few years, as crop yields decline, cattle operations take over the farms and graze the remaining life from the soil. Left behind are barren, eroding landscapes drained of nutrients and incapable of regenerating vegetation.

Throughout this process there is no incentive to replenish the region's resources. As rainforests are logged out and soils lose their productivity, the loggers, farmers, and ranchers all move deeper into the forest in search of virgin land.

Under current conditions, a sustainable use of rainforests is impossible. As long as government policies encourage farming, ranching, and logging without replanting, deforestation will continue. As long as tropical governments refuse to face overpopulation and poverty problems and instead encourage migration into jungle frontiers, rainforest destruction will not stop.

THE UNDERLYING CAUSE

If it is true that imperialists study their colonial charges, it is equally true that the charges study their masters—with great care and cunning. Who shall say which understands the other more?

—Woodrow W. Borah

If rainforests are so important, why are we destroying them? What is the underlying cause?

To answer these questions we must look beyond the local level of small-scale farmers. These people clear rainforests just to survive.

We also must look beyond the national level of tropical governments. These governments encourage logging, ranching, and homesteading in rainforests largely because of financial pressures.

The answer is found on the international level. Specifically, wealth, in the form of both money and natural resources, is being transferred from underdeveloped tropical countries to developed nations outside the tropics.

This transfer of wealth is occurring because many tropical countries are severely indebted to the developed world.

Understanding how this indebtedness came about, how it can never be repaid, and how it dictates policies that encourage deforestation is the key to understanding the driving force behind rainforest destruction.

Origins of a Debt Crisis

How tropical countries became indebted to the developed world is a long and convoluted story. It began in the 1970s when the price of oil started to increase dramatically. Rising oil prices made the countries of OPEC (the Organization of Petroleum Exporting Countries) extremely wealthy.

For safety's sake, OPEC deposited its enormous wealth in the world's most established international banks, which were mostly American and European commercial banks. These banks, in turn, were obliged to invest (lend out) the money, to make a profit, and to pay OPEC interest on its deposits.

Investing the money was difficult, however. There were billions and billions of dollars in OPEC deposits, and the high cost of oil was causing a worldwide recession. This recession made loans to the banks' usual customers in western nations risky.

Several major banks decided that lending to underdeveloped countries was the solution. Many of these underdeveloped countries were in the tropics. Quickly, other, smaller banks followed suit.

The banks considered underdeveloped countries to be like small businesses. The idea was to modernize and industrialize their economies, thereby creating new industries, manufacturing, and products that could be sold worldwide.

Massive amounts of money then flowed to tropical governments and to their industries, most of which were state-owned. It was believed a sovereign nation could not go bankrupt.

THE UNDERLYING CAUSE 41

Where Did the Money Go?

Overall, the many billions of dollars that underdeveloped countries (also called less-developed countries or LDCs) borrowed from the banks was misspent politically. In other words, politicians in these LDCs used the money to stabilize their governments and maintain political power—instead of investing in industries that could yield a profit.

These politicians subsidized food and fuel for their citizens, subsidized the industries of their political supporters, and made government services, such as welfare, education, and police, available to more people.

These underdeveloped countries also spent money buying their own currencies. They did this to strengthen and stabilize their currencies. It was hoped this would spur private investment in their economies and reduce the price of imports, such as machinery and factory equipment that their fledgling industries needed.

The LDCs also needed money to pay interest on their loans. This money was borrowed, as well.

Over time, more and more borrowed money was spent maintaining power, supporting currencies, and making interest payments, and less and less was invested in industry. In the long run, most of the loans meant to build strong economies were misspent through poor political decisions.

Bad Luck Hurt the LDCs

The LDCs did spend some of the money they borrowed to develop their industries, and these industries almost succeeded in strengthening their economies. But around 1980 three changes in the world economy occurred that propelled these emerging nations into a rapid decline—interest rates jumped, oil prices climbed dramatically higher, and prices for international commodities plummeted.

Within four years, from 1977 to 1981, U.S. and other world interest rates increased almost threefold. Because most of their loans were tied to these rates, the LDCs had to make much larger interest payments than they had anticipated.

During this same time, the price of oil skyrocketed to almost $38 per barrel. Consequently, the LDCs needed more money to pay for fuel and the many products whose price is affected by the cost of fuel.

High oil prices also worsened a worldwide economic recession. This, in turn, decreased world demand for commodities like lumber, metals, and farm products. These commodities were the main exports of many LDCs. As demand for these items decreased, the income of these countries also decreased.

All of these events caused the economies of less-developed countries to stumble and the value of their currencies to fall. In the end, high interest rates, high oil prices, less demand for commodities, and the poor spending priorities of the LDCs combined to create huge trade and budget deficits—and a need for more borrowing.

The Brink of Disaster

It quickly became evident that too much money had been lent to the LDCs, too little of which had found its way into economic investment.

Many of these underdeveloped countries were unable to make interest payments on their existing debts but desperately needed more financing to continue industrializing their economies. The banks, however, needed to be paid back and refused to make further loans. By 1982 LDC debt had reached a staggering $780 billion.

The Debt Crisis

At this point, one of the LDCs, Mexico, hinted that it might stop making payments on its foreign debt for a while. Although some small LDCs like Peru and Jamaica had previously made such threats, their debts were insignificant compared to Mexico's. Should a giant like Mexico renege on its financial agreements, other giants were sure to follow.

Soon, some of the smaller underdeveloped countries began discussing a simultaneous default on their loans. These countries believed the banks could not retaliate against whole blocks of bankrupt nations.

The situation rocked the entire international banking system. If a large percentage of less-developed countries were allowed to renege on their debts, banks around the world would be facing insolvency. Financial analysts feared another Great Depression.

Enter the IMF

In desperation, the banks appealed to the International Monetary Fund (IMF) to use its huge financial resources to forestall an international crisis.

The IMF is a financial institution created by the United Nations Monetary and Financial Conference in 1944. It was established to coordinate international payments between countries, promote agreements on currency stabilization, and facilitate international trade.

The IMF gets its money from its 179 member countries. All of these countries, as a condition of membership, must contributed to a pool of money held by the IMF. In return, any one member can borrow from the pool. Large loans, however, require the approval of the other member countries, and strings are often attached.

To ease the situation, the IMF basically offered to lend the LDCs enough money to continue making interest payments to the banks, but only if the LDCs met certain conditions. These conditions included what are called, "structural adjustment programs."

Often controversial, structural adjustment programs require a borrowing country to make changes in its economy, and often these changes benefit the larger members of the IMF while helping the indebted country little.

IMF Advice

The structural adjustment programs of the IMF required the LDCs to alter their economies in two ways: 1) increase exports of natural resources, such as woods, metals, and agricultural products, and 2) cut government spending. In effect, the IMF was telling the LDCs to postpone industrial development and sell off their natural resources.

Many underdeveloped countries were reluctant to make these changes. Prior to IMF involvement, these countries had hoped to repay their debts through increased industrial productivity. But the IMF position was clear: countries that aggressively followed its advice would have their past-due payments refinanced, receive favorable repayment schedules, and be given more loans in the future.

Not wanting their countries to be isolated from the global economic system and needing more money to stay in power, most LDC officials accepted the new loans and followed the IMF advice.

THE UNDERLYING CAUSE 45

The Debt Crisis Begins to Fuel Deforestation

Exporting Rainforests

The first part of the IMF plan—to increase exports of natural resources—was accomplished in several ways. One way was to require the LDCs to expand their industries that produced commodities for export. Consequently, local companies involved in industries like timber, agriculture, cattle, and mining received government subsidies.

The IMF also insisted that the LDCs lower trade barriers, such as taxes, tariffs, and restrictions on foreign firms. With low trade barriers, international companies could help the LDCs export natural resources. With modern equipment and direct access to foreign markets, these large, international businesses began extracting resources on a massive scale. (In some cases, these companies were financially larger than the countries within which they operated.)

Finally, the IMF required the LDCs to stop expending money to buy their own currencies. As a result, the value of these currencies plummeted, and the price of everything—labor, materials, land—became less expensive to foreign firms. Consequently, the exporting of natural resources by foreign firms became extremely profitable.

All of these policies were endorsed by the developed world in the name of "free trade." In the underdeveloped countries of the tropics, however, the very industries that benefited from IMF policies are the industries that, today, destroy rainforests.

Creating Poverty

The second part of the IMF repayment plan—to cut government spending—also was endorsed by the developed world. However, these cuts in spending increased poverty in these already poor countries, and this eventually caused more deforestation.

The IMF believed that spending cuts would reform LDC governments, end wasteful spending, and save money to repay debt. But to the people of less-developed countries, these spending cuts had tremendous social consequences and were called "austerity policies."

For LDC citizens, austerity policies meant higher prices for food and fuel, fewer educational opportunities, wage and job cuts by the government and by state-owned industries, cutbacks in welfare and social security programs, and the elimination of many basic city services—in other words, recession.

Also contributing to poverty in these countries was insistence by the IMF that less money be spent supporting currencies. As the value of currencies fell, everything these small countries imported became drastically more expensive.

Austerity Backfires

Suddenly, LDC residents were experiencing widespread poverty. Food was in short supply and jobs could not be found. Citizens were upset that their governments had acquiesced to the policies of foreign nations and they organized strikes, riots, and demonstrations against their governments, the IMF, and the largest shareholder of the IMF, the United States.

Between 1976 and 1989 these "austerity protests" were reported in 26 different debtor countries.

One of the most violent protests occurred in Venezuela in March 1989. There, outraged Venezuelans burned cars, looted stores, and fought police, resulting in the deaths

THE UNDERLYING CAUSE 47

of 300 people, injuries to 2,000, and strong reprisals from the Venezuelan government.

The *Economist* reported on the response of then-Venezuelan President Carlos Andres Perez: "Mr. Perez responded to the riots by ordering a curfew, suspending the right of assembly and of free speech, and permitting detention without trial. He also strongly defended the austerity measures against which the rioters were protesting, and upon whose implementation the International Monetary Fund has insisted as the condition for the first of what may be a long line of fresh credits to Venezuela."

In another country, the Philippines, austerity protests eventually toppled the U.S.-backed Marcos regime.

The LDC Response

Government officials in these underdeveloped countries were in an awkward position. They needed to quiet political unrest, and they genuinely wanted to provide for their people, many of whom were struggling to survive. Their response, however, dramatically increased deforestation by creating millions of small-scale farmers.

To LDC government officials, the best way to deal with impoverished and discontent citizens was to offer free or nearly free land in the countryside where these people could start their own farms and grow their own food.

In tropical countries, however, good agricultural lands were needed to raise export crops—to repay foreign creditors. The only land for the poor was in the rainforest frontier.

Rainforest Colonization

As discussed in chapter three, the LDCs of the tropics did a number of things to encourage their citizens to move into rainforests: they built highways into remote jungle areas, made land grants to citizens who would settle there, and made loans and subsidies available to those settlers who wished to start a farm or ranch in these areas.

The Role of the World Bank

Much of the financing for these rainforest development projects came from international agencies, such as the World Bank.

The World Bank is the sister organization of the IMF. Like the IMF, it is an intergovernmental organization created by the United Nations Monetary and Financial Conference of 1944. Its primary function is to make loans available to low-income countries for specific development projects, such as dams and highways. The World Bank also provides economic advice and technical assistance to underdeveloped countries.

Resettlement Programs Financed by the World Bank

Between 1986 and 1993 the World bank approved 146 projects that involved the resettlement of people—15 percent of the World Bank's budget. Many of these resettlement programs involved moving people from crowded urban centers into sparsely populated rainforests

In Indonesia, for example, over $500 million in World Bank financing helped colonize that country's perimeter islands. The gist of this program was to relocate people from the overpopulated island of Java to remote islands, which were covered in rainforests.

Indonesia's colonization effort continues today and is the largest government-sponsored program encouraging

THE UNDERLYING CAUSE 49

people to move into rainforests. It is estimated that 10 million people have been resettled under this program.

The World Bank also provided Brazil's Polonoroeste Regional Development Program with more than $400 million. Polonoroeste expanded Brazil's highway 364 into the rainforests of Rondonia, Brazil, subdivided land, and gave land titles to settlers.

The result of Polonoroeste was a land-rush and a wholesale leveling of rainforests. During most of the 1980s, over 100,000 people per year moved into the rainforests of Rondonia, most because of Polonoroeste. In 1984 alone, nearly 140,000 new settlers migrated into this rural area.

Land titles, however, were unavailable for so many people. With no other way to survive and nowhere else to go, most of these settlers have become squatters—clearing and farming rainforests they do not own. From 1980 to 1990 forest cover in Rondonia dropped over 20 percent.

Today, an estimated 300 to 500 million small-scale farmers in similar predicaments are clearing rainforests worldwide.

What is the Situation Now?

The IMF and other intergovernmental institutions still have policies that cause deforestation. These policies also increase the debts of less-developed countries; the commercial banks, however, have been rescued.

The Banks Escape
Since the IMF intercession, most commercial banks threatened by the debt crisis are no longer facing insolvency.

The IMF bought the time these banks needed to reorganize themselves. Some of these banks built up cash reserves in anticipation of future losses. Others got rid of their troubled loans, selling them at discounts to other banks. Still others negotiated guarantees for their loans from institutions like the World Bank.

In addition, many banks are being paid back—largely from the sale of natural resources. By the late 1980s, nearly 30 percent of all the money that less-developed countries earned from exports went to pay debt; in Latin America it was nearly 40 percent.

As a consequence, the 1980s saw a net flow of money from underdeveloped countries to foreign creditors. In other words, the LDCs paid more to their foreign creditors in the 1980s than they received in additional loans and foreign investments. By 1989 $50 billion per year was flowing out of less-developed countries.

For poor countries, the loss of this money is devastating. It means fewer dollars for economic investment, social programs, and environmental protection. In other words, more poverty and more deforestation.

The Plight of the LDCs

Despite the flow of money from less-developed countries to foreign creditors, LDC debt has increased. In fact, it has doubled. Much of this debt is refinanced interest, however. The debts of less-developed countries now total $1.4 trillion—about twice what it was in 1982 when the IMF interceded.

As a result, many LDCs are trapped in a vicious cycle. Their debts are increasing as interest accumulates, and their supply of natural resources, which they sell to repay these debts, is shrinking. If current trends continue, many LDCs soon will be stripped of natural resources but still saddled with massive debts.

COLONIALISM?

To people sympathetic to the plight of underdeveloped countries, the present situation is a modern-day version of colonialism. The IMF and other financial institutions are promoting policies that benefit the developed world at the expense of less-developed countries.

In its 1994 annual report, for example, the IMF stated that "improved prospects for these (underdeveloped) countries would require strong adjustment programs," including a reduction of protectionist policies and a reduction of exchange rates. These policies, however, will make exporting natural resources from less-developed countries more profitable for international companies. As a result, more rainforests will be destroyed.

Of course, underdeveloped countries do not have to follow IMF advice; they can decline to restructure their economies around the export of natural resources. But doing so is not politically feasible.

This is because most of the land in these countries is owned by a small group of politically powerful people. In tropical Latin America, for example, 90 percent of the arable land is owned by just seven percent of the population.

These politically powerful land owners profit from the sale of their countries' natural resources. They have little incentive to change existing policies.

To the poor and politically powerless in less-developed countries, the current situation is colonialism. Their countries are used by international corporations with the consent of a wealthy political elite to produce goods for foreign markets. Foreign firms reap the profits from these products, and tax revenues that could otherwise be used to develop diversified economies are used instead to pay foreign debts.

Colonialism or not, one thing is clear—the poor resolution of the debt crisis is the driving force behind rainforest destruction. The developed world's financial centers are pressuring tropical nations to sell natural resources and ignore long-term social and environmental consequences.

This pressure is what drives the non-sustainable use of resources in the tropics. As long as it remains, rainforests will be destroyed.

THE WORLD ON FIRE

No Previous generation has had to think of, much less respond to, this phenomenon.

-Jay D. Hair
National Wildlife Federation

The following photos were taken by space shuttle astronauts and were provided courtesy of NASA. They show the smoke plumes of farmers and ranchers burning vegetation in the countries of Indonesia, Argentina, Mozambique, Brazil, Bolivia, Papua New Guinea, Zimbabwe, and Tanzania.

The first photo is perhaps the most disturbing. It was taken by Discovery astronauts over the Amazon and was released from NASA with the following caption.

"South America's Amazon basin is obscured by smoke from the clearing and burning of tropical forest....The smoke cloud observed during this mission was the largest and densest yet seen by astronauts. If placed over the United States, it would have covered an area of the country more than three times the size of Texas. This view shows a portion of the smoke ending where the Andes mountains rise on the horizon. The closest mountains are about 650 miles distant."

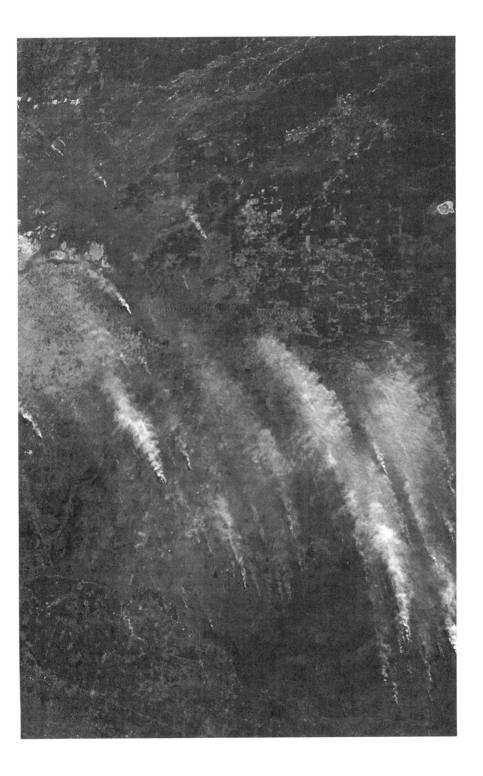

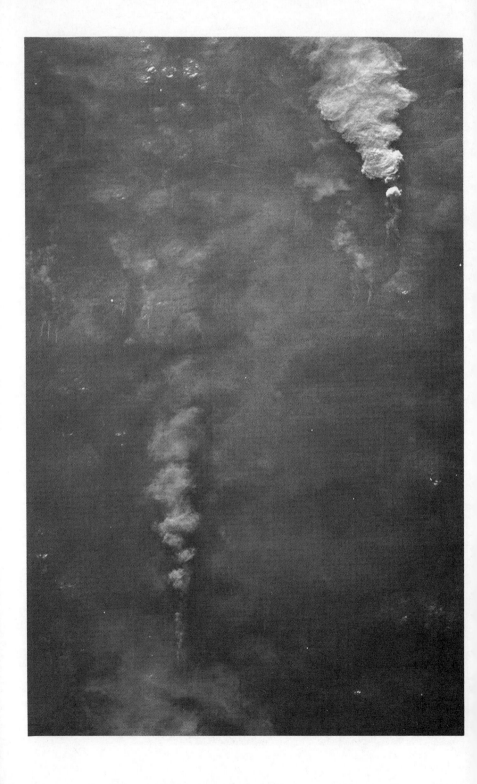

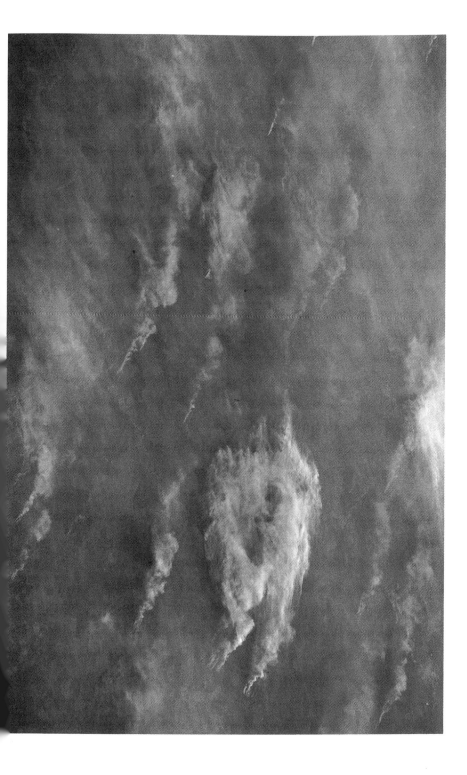

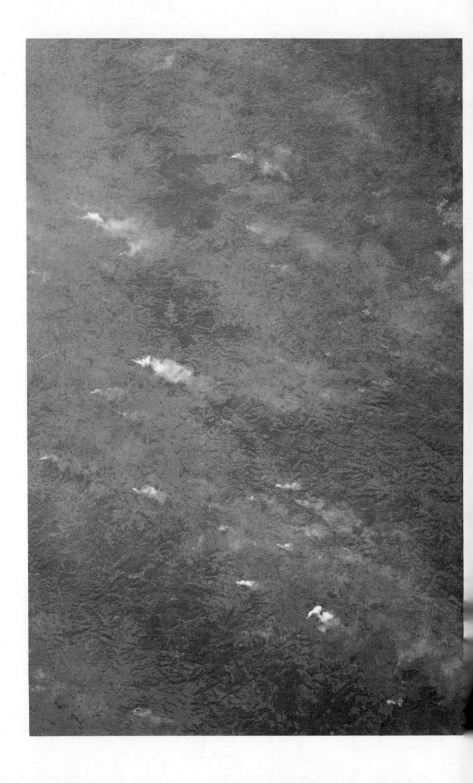

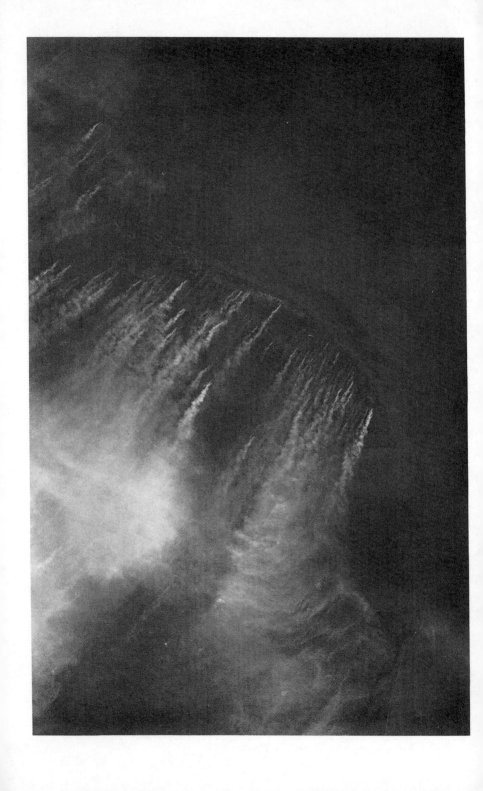

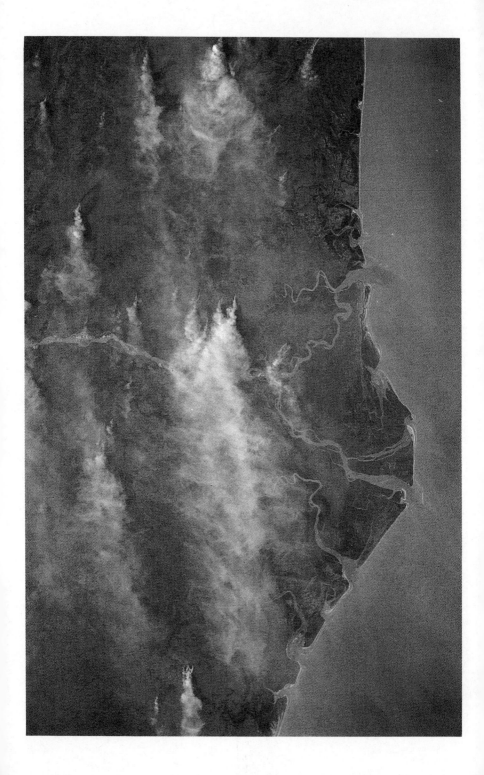

A CONFLICT OF CULTURES

Where today are the Pequot? Where are the Narragansett, the Mohican, the Pokanoket, and many other once powerful tribes of our people? They have vanished before the avarice and the oppression of the White Man, as snow before a summer sun.

—Tecumseh of the Shawnee 1813

In the late 1700s an Indian chief named Tecumseh lived along the East Coast of the North American continent. In his lifetime he witnessed the defeat and disappearance of numerous Indian tribes as the region was deforested and settled by European descendants.

Today, as unspoiled rainforest regions are deforested and settled by outsiders, primitive Indian tribes again are being extinguished. History is repeating itself.

Tecumseh

Tecumseh belonged to a tribe called the Shawnee. In its formative years, the United States forced the Shawnee and many other eastern Indian tribes to move to land west of the Mississippi River.

Tecumseh, however, would not move west. His father and brother had been killed by American troops and he was determined to resist the American conquest of his tribe's former home.

Tecumseh instead traveled to regions inhabited by other tribes. He urged these tribes to form an alliance to stop the expansion of the United States. Tecumseh's travels were so extensive that it is believed he contacted every Indian tribe between the Atlantic coast and the Rocky Mountains.

By the early 1800s, Tecumseh had assembled a following of about 800 warriors from the tribes he had contacted. His forces, however, were ill-equipped to fight the U.S. Army. In 1811, at the Battle of Tippecanoe they were defeated.

Despite this defeat, Tecumseh and what remained of his followers continued to resist the Americans. In the War of 1812, they joined the British to fight against the United States. In this conflict, at the Battle of Thames River, Tecumseh was killed.

Because of his effort to build an Indian confederacy and protect his homeland, Tecumseh is regarded today as one of the greatest Indians who ever lived.

Andrew Jackson

During the time of Tecumseh, Andrew Jackson, who would later be elected the 7th president of the United States, became famous for fighting against the Indians.

As a young man, Jackson had an illustrious military career, serving as a major general in both the Tennessee militia and the United States Army. During this time, Jackson helped kill thousands of Cherokees, Chickasaws, Choctaws, Creeks, and Seminoles.

Years later, as President of the United States, Jackson continued his crusade against the Indians. In his first address to Congress, Jackson spoke about his plans for the Indians. His remarks underscore the connection between native cultures, natural resources, and colonial powers.

A CONFLICT OF CULTURES 71

> *Our ancestors found them (the Indians) the uncontrolled possessors of these vast regions. By persuasion and force, they have been made to retire from river to river, and from mountain to mountain; until some of the tribes have become extinct, and others have left but remnants....Surrounded by the Whites, with their arts of civilization, which by destroying the resources of the savage, doom him to weakness and decay; the fate of the Mohegan, the Narragansett, and the Delaware, is fast overtaking the Choctaw, the Cherokee, and the Creek....I suggest the propriety of setting apart an ample district west of the Mississippi, and without the limits of any State or Territory, now formed, to be guaranteed to the Indian tribes, as long as they shall occupy it.*

To Jackson, relocating the Indians west of the Mississippi River must have seemed the best solution to the problems facing his young nation. In June of 1834, during Jackson's second term as president, Congress concurred and passed *An Act to Regulate Trade and Intercourse with the Indian Tribes and to Preserve Peace on the Frontiers.*

This Act decreed that all land west of the Mississippi River was to be Indian Country except the existing states and territories of Louisiana, Missouri, and Arkansas. This solution, however, did little to quench the United States' thirst for resources.

Thirsting for Resources

Within the next 80 years, the United States expanded westward into 48 states. Within these states burgeoning agricultural and timber industries deforested some 400 million acres of Indian land—an area twice the size of Texas.

As this annexation and development of Indian land occurred, the rights of native Americans were systematically taken away: the U.S. Army forced Indians onto reservations; the U.S. Supreme Court declared that an American

Indian did not have the right to vote; and the U.S. Congress passed laws that sold Indian land to U.S. homesteaders. (In fact, the sale of Indian land raised enough money to pay off the national debt.)

Also during this time, successive westward waves of American and European immigrants, the diseases they brought, and the United States Army decimated the population of Native Americans to less than one-third its pre-colonial number.

Repeating History

The price Native Americans paid for the progress of the United States is well known. Less well known, however, is that today Indian tribes in tropical rainforests are besieged by similar colonial powers.

Now, as in the formative years of the United States, some of these Indians are fighting to preserve their ancient cultures and their primeval forests.

What follows is an examination of four such tribes, their beliefs, and their battles to resist incursions on their land.

THE PENAN BATTLE INTERNATIONAL LOGGING COMPANIES

The Penan Indians are one of the world's last nomadic tribes of hunter-gatherers. They live on the island of Borneo in the Malaysian state of Sarawak. The largest of 14 Malaysian states, Sarawak has about 34,200 square miles of rainforest, all of which is state-owned.

The Penan Way of Life

The Penan lead a nomadic life that does not damage the environment. They do not clear land for farms; all that the tribe needs is produced naturally by the rainforest, as it has been for centuries.

The Penan are a primitive people. Visually striking, they still dress in loincloths, hunt with blow-pipes, and decorate themselves with huge, circular earrings that stretch their earlobes to their shoulders.

The Penan have a superstitious respect for their environment. They believe that their rainforest is full of spirits. They refuse to cut down large trees or take more than they need from the forest for fear of angering these spirits.

The Penan also have beliefs that keep them nomadic. Certain bird calls and other animal signs are considered omens to move from place to place. Whenever a member of the tribe dies, the entire village relocates to another part of the rainforest. These beliefs keep the tribe from overharvesting the forest in any one area.

To outsiders, Penan superstitions are not meaningful. But for the Penan, their beliefs are what protect their environment.

Changing Times

Several decades ago, the modern world began to undermine the nomadic and spiritual ways of the Penan.

Christian missionaries were the first to influence the tribe. Less than fifty years ago, these missionaries urged the Penan to cut down large trees, build permanent houses, and remain settled despite the misgivings of tribal elders.

In the 1970s the Malaysian government also tried to "civilize" the tribe by sending Penan children to boarding schools. Attending these schools, however, separated the children from their parents and the lessons and traditions of their culture.

Today, the last of the Penan forest culture is threatened by big business and deforestation. In the past two decades, international lumber companies have created a huge logging industry in Sarawak. This industry is so big that it now accounts for almost half of Sarawak's total state revenues and about half the exports of tropical logs worldwide.

As a result, deforestation rates in Sarawak are among the highest in the world. In 1991, the logging rate for this region was estimated at 850,000 to 1,000,000 acres a year—almost two acres a minute.

This logging of Sarawak is destroying the rainforest where the Penan live. Today, only a fraction of the tribe can still survive by hunting and gathering in the rainforest. Most Penan have had to switch to subsistence farming, abandoning their traditional culture.

Tribal Resistance

The Penan have tried to fight Sarawak's timber industry, but with limited success. In 1987 the tribe barricaded logging roads with tree branches. When these efforts failed to stop the loggers, the Penan used themselves as human barricades to protect their rainforests.

The Sarawak police then arrested a number of the Penan. The road blocks, however, did not stop.

Soon, the attention of the Western news media was captured as were the hearts of some Western leaders. In a February 1990 speech, England's Prince Charles condemned Malaysia's treatment of the Penan, saying, "The Penan in Sarawak are hassled and even imprisoned for defending their own tribal lands. Even now, that dreadful pattern of collective genocide continues."

Sympathetic Europeans began to boycott products made with Malaysian timber. To help with these boycotts, the Austrian government passed a law requiring tropical wood products labeled "made of tropical wood."

Boycotts in Europe had little impact, however, because almost half of Sarawak's lumber is exported to Japan. After an investigative trip to both Sarawak and Japan in 1991, the *New Yorker* reporter Stan Sesser wrote, "The voracious Japanese appetite for tropical hardwood has turned Sarawak into something resembling a Japanese plantation."

Sesser described the effects of this deforestation on the few remaining Penan who refuse to abandon their hunting culture, "The hunters had walked for miles and had been without food for the entire day, but when I glanced at their faces I saw more than hunger. A pall hung over them—the look of men who could no longer provide for their families."

Despite the work of Sesser and the efforts of European boycotters, it is unlikely that the Penan culture will survive much longer. The tropical timber industry's own trade association, called the International Tropical Timber Organization, estimates that Sarawak's remaining undisturbed forests will be logged out by 2001. Once this happens, the last of the Penan hunting and gathering culture will be gone.

The Yanomami and the Gold Rush of the 1980s

Another rainforest Indian tribe facing development pressures is the Yanomami of Brazil. The Yanomami, however, are not only experiencing a cultural change, they are facing extinction.

The Yanomami

Like the Penan, the Yanomami are a primitive tribe of hunter-gatherers. They wear little clothing, preferring instead to adorn themselves with feathers, flowers, and skin dyes.

The Yanomami share with other Indian tribes a remote 60,000-square-mile rainforest habitat along the Venezuelan and Brazilian border in the Parima Mountains. It is estimated that the Yanomami number about 25,000 people, but the size of their tribe is decreasing rapidly.

Yanomami Society

The Yanomami tribe consists of roughly 200 separate villages, each of which relocates every few years to allow the surrounding forest to regenerate. Each village consists of several large, round, thatched huts within which several Yanomami families live communally.

Anthropologists consider the Yanomami tribe to be one of the world's most violent cultures. Disagreements often are settled by fights to the death.

Yanomami conflicts differ, however, from violence in the developed world. A recent study by Napoleon Chagnon, an anthropologist from the University of California at Santa Barbara, reveals that the Yanomami are polygamous and fight, not for possessions, land, or resources, but for wives.

Disease Invades the Yanomami Territory

Despite the fierceness of the Yanomami, outsiders have not been deterred from invading the territory of these Indians. When gold was discovered on Yanomami land in the 1980s, tens of thousands of prospectors and panhandlers moved in.

A CONFLICT OF CULTURES 77

These miners far outnumber the Yanomami. According to *Science* magazine, "As many as 70,000 gold prospectors have invaded an area in northern Brazil traditionally occupied by 7,500 Yanomami. More miners, poor but dreaming of gold, come every day."

Contracting diseases from these prospectors is the greatest threat to the Yanomami. Living in virtual isolation until the 1980s, the tribe has developed little natural immunity against many viruses. Simple infections, even colds, can kill these Indians.

Previous contact with outsiders has been limited to the area's few missionaries. According to Charles Brewer-Carias, a naturalist working to establish a Yanomami reserve in Venezuela, even these missionaries have given fatal diseases to the Yanomami.

Carias wants to establish a preserve for the Yanomami that is off-limits not only to merchants, prospectors, and tourists, but also to missionaries. Carias realizes that this is considered a radical idea in a predominantly Roman-Catholic country like Venezuela, but he observes, "Our religion has no place in their lives. The Yanomami have nothing to be saved from but ourselves."

Government Action
Environmental and indigenous rights organizations have demanded that Brazil protect the Yanomami. However, the Brazilian government also is interested in mining Yanomami lands, not only for gold but also for diamonds, titanium, tin, and uranium.

The rainforests along Brazil's northern border, which include the Yanomami lands, are thought to contain enough mineral resources to pay off almost all of Brazil's $120 billion foreign debt.

To protect these resources, Brazil's military is establishing outposts in the region. The official plan is called *Calha Norte* and eventually will deforest much of the area.

Calha Norte is, in effect, a security zone 4,000 miles long and 93 miles wide between Brazil and its neighbors: Peru, Colombia, Venezuela, and the three Guianas. The plan's ultimate goal is to establish a self-sufficient region connected with roads, communications, and power systems.

The first part of the plan was started in 1986 with the building of air fields in the jungle. Unfortunately for the Yanomami, these airfields only brought in more miners.

Within four years after the completion of these airfields, 15 percent of the Brazilian Yanomami had died from outside diseases. In 1991 *Science* magazine reported the annual Yanomami death rate to be 13 percent and the fertility rate near zero. Some Yanomami villages are suffering a 90-percent sickness rate.

If the Yanomami are to be saved, the Brazilian government needs to quickly preserve a territory for the tribe and Carias's Venezuelan plan must be implemented soon. According to the American Anthropological Association, the Yanomami, without protection, will be extinct within the next decade.

Environmental Organizations on the War-Path

It appears that the Penan and the Yanomami have little chance for survival. But there are some tribes, like the Kuna of Panama and the Kayapo of Brazil, whose rainforest and sustainable way of life are being protected with the help of environmental organizations.

A CONFLICT OF CULTURES

These environmental organizations include, among many others, the World Wildlife Fund, the National Wildlife Federation, Cultural Survival, and the Environmental Defense Fund. Although generally concerned with environmental issues, these organizations realize that by protecting indigenous cultures, they also are preserving the earth's natural resources.

THE KAYAPO STOP THE WORLD'S LARGEST DAM PROJECT

For many centuries an Indian tribe called the Kayapo has fought to defend its rainforest homeland along the Xingu tributary of the Amazon River. In the 17th through 19th centuries this tribe battled Portuguese slave traders and gold prospectors in the rainforest.

In the 20th century, however, the Kayapo have existed peacefully with the outside world. But, this was before the outside world threatened to completely destroy their rainforest homeland.

The Kayapo Indians

The Kayapo are a tribe of about 3,000 people. They live in 13 known villages scattered throughout hundreds of miles of Brazilian rainforest, an area they share with a number of other Indian tribes.

The Kayapo are a statuesque people, standing over 6 feet tall. Bands of beads adorn their necks and arms, and both men and women wear large feathered earrings. On ceremonial occasions these Indians dress in feathered headdresses and stretch their lower lips outward over large wooden disks.

Kayapo Knowledge of the Rainforest

The Kayapo are not a nomadic people, however, they do practice a unique "nomadic agriculture." Their farms are only a few feet wide and meander for miles along trails in the rainforest.

This forest-agriculture is, in effect, the opposite of Western agricultural practices; there is no clearing of fields or environmental damage, just a few plants sown every few feet along either side of a rainforest trail.

The Kayapo also possess a unique knowledge of rainforest plants. This knowledge has been studied by American ethnobiologist Darrell Posey in the Kayapo village of Gorotire for over twelve years.

Posey's work reveals that the Kayapo, in addition to knowing what is edible and what is poisonous, use some plants as natural insecticides and others as natural herbicides. A number of plants have medicinal purposes. Some even prevent pregnancy.

Government Policy

As is the case with the Yanomami, the Kayapo way of life is not valued by Brazil. It is government policy to assimilate all Indians into the country's larger society.

Brazil, however, has designated a reserve for the Kayapo. But its boundaries are not well marked nor permanent.

Electronorte

Recently, the Brazilian power company, Electronorte, tried to alter the boundaries of the Kayapo reservation to build a series of hydroelectric dams. If completed these dams would have submerged enormous areas of rainforest, much of which belongs to the Kayapo and other Indian tribes.

A CONFLICT OF CULTURES

The Electronorte project was to be Called the Xingu River Complex and was to be financed by the World Bank. In all, the project would have included 47 dams along the Xingu River and would have created the world's largest man-made body of water.

Two of the dams would have required the relocation of several Kayapo villages. Five of the dams would have created a reservoir covering 6,800 square miles of rainforest—an area almost as big as New Jersey.

No one knows exactly how much flooding would have resulted had the entire project been completed. But to the Kayapo, no amount of electricity was worth the loss of their homeland.

The Kayapo Fight Back with Modern Technology

One of the Kayapo chiefs, Paulinho Paiakan, decided to fight the project.

For many years Paiakan had served the Kayapo as a liaison to the outside world. Because of this experience, Paiakan was acquainted with a number of environmental and indigenous-rights organizations and familiar with modern-day technology.

Paiakan convinced several other Kayapo chiefs that, with the help of these outside groups and the use of modern technology, they could save their rainforest by exposing Electronorte's plan as an environmental disaster.

The first thing Paiakan needed to do was educate and unite his tribe. He and the other chiefs did this by touring and filming with camcorders a recently built dam on the Tocantins River. This dam's reservoir had submerged some 800 square miles of rainforest—the homelands of the Parakanan and Gavioes Indians.

The chiefs then transferred their films onto video tapes. Using gasoline-powered generators, VCRs, and portable TVs (which the Kayapo call the Big Ghost), the chiefs showed the video tapes in the scattered Kayapo villages.

The Kayapo understood. They discussed the situation and agreed that the best way to expose the Electronorte project was to protest the largest proposed dam near the town of Altamira.

With telephones and modern-day fax machines, they notified the world press and environmental groups of their plans.

Then, in February 1989, 500 Kayapo converged on Altamira in full ceremonial dress.

With their war-clubs in hand, the 500 Kayapo emerged from the forest and surrounded a meeting of Electronorte officials. Before the watchful eyes of the media, the tribe staged a dramatic but peaceful demonstration. They danced tribal dances, performed ancient rituals, and made speeches denouncing Electronorte's flooding of rainforests. The Kayapo protest received worldwide media coverage.

A Trip to Washington

The National Wildlife Federation and the Environmental Defense Fund also focused attention on Electronorte's proposed Xingu River Complex. These environmental organizations brought Chief Paiakan and another Kayapo leader, Chief Kube-i, to Washington D.C. to lobby World Bank officials not to fund the project.

The Kayapo chiefs, accompanied by their ethnobiologist friend Darrell Posey, also met with members of the U.S. Congress and officials from the U.S. state and treasury departments.

Under pressure from the international environmental community, the World Bank withheld the project's initial $500 million funding. Without funding, the project was canceled. The Xingu River Complex and the rainforest destruction it would have caused had received too much attention.

Pressure to develop the Amazon, however, remains. Electronorte still has plans to build 11 other dams in the region, which, if completed, will submerge 3,800 square miles of rainforest.

The Kuna Save Their Reserve From Slash and Burn

Large-scale development projects and international demand for natural resources are not the only threats to the rainforest and its Indian inhabitants. Recently, the Kuna, a tribe of Panamanian Indians, found their reservation threatened by scores of small-scale farmers.

For the most part, these small-scale farmers are the result of exploding urban populations. In search of a better life, they move into rainforests along jungle roads built by their governments or by logging companies. Once in the jungle, they settle on land they do not own, clearing small plots in order to grow crops or raise cattle.

The environmental damage that each individual farmer does is relatively small, but cumulatively the damage of many such farmers is tremendous.

The Kuna Tribe

The Kuna Indians of Panama are among the world's shortest people; most are no taller than 5 feet. The Kuna number about 30,000 and are the largest Indian tribe in the American tropics. Typical dress for the Kuna includes beads, gold jewelry, and hand-made clothes that are brightly colored.

Kuna Isolation

The Kuna possess an 1,800-square-mile rainforest reservation between Panama's Caribbean coast and the San Blas Mountains to the southwest. Most Kuna live on small coral islands located just off the coast but travel daily by canoe to their mainland reservation for farming, hunting, and gathering.

For the Kuna, there is no reason to store produce or interact with the outside world. Their jungle and coastal area provide food and supplies year-round. Their reservation is completely isolated. No roads lead to it. The only access is by boat.

An Army of Small-scale Farmers

The Kuna have legal title to their reservation. Panamanian law forbids any non-Kuna from owning land or squatting in the tribe's territory. But because Panamanian police cannot patrol the entire rainforest, these laws are not enforced.

Consequently, the Kuna became concerned when a feeder road from the Pan-American Highway, called the El Llano-Carti road, was being built through the jungle toward their isolated reserve. Already, small-scale farmers were burning the rainforest and establishing farms illegally along the completed portions of the El Llano-Carti road. It was only a matter of time until the road and the migrant farmers reached the Kuna reservation.

A CONFLICT OF CULTURES

A Defensive Strategy

The Kuna reviewed the situation with forestry experts from the Tropical Agronomic Center for Research and Training in Costa Rica. All agreed that the building of the El Llano-Carti road could not be stopped. The Kuna's only defense was to develop their land before the squatters arrived.

So, with technical and financial assistance from various international organizations, including the Intra-American Foundation and the World Wildlife Fund-U.S., the tribe created a 230-square-mile botanical park on the edge of their reserve.

The Kuna strategically situated the park at the very point where the new road would intersect their land. The park's perimeter is clearly marked and constantly patrolled by Kuna rangers. No one gets in or out without Kuna permission.

As part of the park, the Kuna built housing for tourists and scientists. Revenues from these facilities pay the park's expenses. The park is now constantly occupied. In the vastness of the jungle, occupying land is often what determines legal ownership.

As another protection, the Kuna want Panama and the United Nations to declare their park a "biosphere reserve."

A biosphere reserve is an area of land voluntarily set aside by a country and monitored for environmental protection by the United Nations Educational, Scientific, and Cultural Organization (UNESCO). To date, UNESCO monitors 300 biosphere reserves in 76 different countries, protecting 600,000 square miles of wildlife habitat.

The Kuna believe that if their park becomes a biosphere reserve it will be better protected; any incursions on their land would then receive international attention. But even without this designation, the Kuna's park has protected their rainforest from the most pervasive deforestor, the slash-and-burn farmer.

Remaining Tribes

The Kuna, Yanomami, Kayapo, and Penan represent only a fraction of the world's remaining rainforest Indians. Their struggles, however, are representative of a larger effort worldwide to preserve Indian cultures and the environment.

The plight of these four tribes is also a modern-day version of the battles that North American Indians fought and lost a century ago on the frontiers of North America.

The settlement and development of Indian lands by foreign cultures is not just a phenomenon of the past. In this century, incursions into the Amazon region alone have resulted in the loss of over 80 tribes, and many of the surviving tribes have dwindled to fewer than 1,000 people. Unless something is done, the world's remaining forest cultures will disappear in our lifetime.

NEW DIRECTIONS

The world we have created today as the result of our thinking thus far has problems which cannot be solved by thinking the way we thought when we created them.

—Albert Einstein

The truth is, we are all part of a culture that supports itself through the non-sustainable use of resources. Our economy is such that we prosper at the expense of the environment. Our way of life depends on deferring the costs of environmental damage to future generations and to people in other countries.

Such a culture cannot last. Eventually, we will destroy our environment and our civilization. However, a new environmentalism has emerged that would recognize environmental costs today, force us to pay as we go, and bring us back from the ecological Rubicon of mass extinction which we are about to cross.

This new environmentalism incorporates ideas of sustainability into our modern economy. It envisions a world where natural resources are consumed but also replenished, where rainforests are inhabited and yet flourish. This new environmentalism is called "sustainable development."

Defining Sustainable Development

For centuries, the philosophy behind sustainable development—to live in harmony with the environment—has been an integral part of most indigenous cultures.

The term "sustainable development," however, was first coined in 1987 in a United Nations' report entitled, *Our Common Future*. This report defined sustainable development as "development that meets the needs of the present without compromising the ability of future generations to meet their own needs." In other words, as natural resources are used, they are replaced.

Sustainable Development and Rainforest Destruction

Sustainable development has implications for almost every environmental problem. With regard to tropical deforestation, however, there are five ways sustainable development can alter the fate of our rainforests.

These are: 1) create rainforest preserves that will pay for themselves, 2) support earth-friendly businesses, 3) use governmental incentives to require companies to replace the natural resources they use, 4) change the policies of intergovernmental lending institutions that currently finance environmentally destructive economies, and 5) reduce the debts of tropical countries to a manageable level.

The next few pages will examine each of these sustainable development solutions and identify some of the dedicated men, women, and organizations working to integrate them into our global economy.

Rainforest Preserves that Profit

Establishing rainforest parks and preserves can help save rainforests. However, setting aside natural areas costs money. In underdeveloped tropical countries, funds for such projects often are not available.

In such cases rainforest preserves are needed that can pay for themselves. This can be done through "ecotourism" or "extractive reserves."

What is Ecotourism?

Ecotourism is the practice of charging tourists a fee to visit a park or preserve and then using these fees to protect the natural fauna of the area. It is, in essence, a user fee charged to tourists and earmarked to protect nature.

Throughout the world there are about 5,400 protected natural areas generating an income from tourism. But so far only about 5 percent of the world's forests are in these areas. More preserves built around rainforests and more income from these preserves are needed to help save rainforests.

A good example of rainforest ecotourism is in Madagascar. There, tourists are paying to see the rare, raccoon-like Lemur in his native forest. Another example is in Brazil, where the largest monkey, the Muriqui, has become a tourist attraction. Before the civil unrest in Rwanda, however, one of the best examples of rainforest ecotourism was that country's Park National des Volcans.

Rwanda and the Mountain Gorilla

Created more than 60 years ago, the Park National des Volcans was a huge wildlife sanctuary that protected the endangered Mountain Gorilla and other wildlife of central Africa. But in the early 1970s, because of an expanding population and poverty problems, the Rwandan government converted half the park to farmland and was debating the fate of the remaining half. Many environmentalists feared the park would soon disappear.

In 1979 two environmentalists, Amy Vedder and her husband Bill Weber, convinced the Rwandan government that endangered species tourism could be a profitable use for the park. Vedder, a zoologist with the New York Zoological Society, and Webber, a tropical research specialist, envisioned tourists paying to see, in a natural habitat, the endangered gorillas made famous by Dian Fossey's book *Gorillas in the Mist*.

The two then set about acclimating the normally shy, but sometimes aggressive, gorillas to human visitors. Before the Rwandan civil war erupted in 1994, about 18 tourists a day were paying for a one-hour guided tour among the gorillas. Because of these tourists, park revenues increased by 2,000 percent, and the Rwandan government decided to maintain the park as a tourist attraction.

Extractive Reserves

Rainforests also can generate profits—and thus ensure their own protection—through the creation of extractive reserves. An extractive reserve is an area in the rainforest set aside for the harvesting of crops that occur naturally in the environment, such as herbs, fruits, nuts, and rubber.

The idea behind an extractive reserve is to harvest only what the forest can easily regenerate. In this way, an extractive reserve becomes a perpetual source of income and thus a perpetual means of protecting rainforests.

In the long run, the income from an extractive reserve should exceed the income from a non-sustainable use of a rainforest, such as clear-cut logging. Charles Peters, a botanist with the New York Botanical Garden, and two associates, Alwyn Gentry and Robert Mendelsohn, proved this to be true in the Peruvian Amazon.

These three men acquired a one-hectare plot of rainforest (2.5 acres) near the town of Mishana, Peru. There, they inventoried all the crops that could be harvested seasonally—fruits, vegetables, wild chocolate, latex, etc. They estimated the value of these crops to be about $420 at a local Peruvian market.

Peters, Gentry, and Mendelsohn then estimated the value of the timber on their land. They found that a local saw mill would pay them $1,000 if they clear-cut the entire plot. They concluded that if left forested and harvested repeatedly for sustainable crops, the rainforest in this area would, over time, produce considerably more income than clear-cutting for timber one time.

Supporting Sustainable Businesses

Everyone of us, as a consumer, can help preserve rainforests. We can do this by demanding goods made with concern for the environment, including products from extractive reserves. Businesses that produce these earth-friendly products have been dubbed "sustainable businesses."

Ben and Jerry's
A natural ice cream company, Ben & Jerry's Homemade Inc., is one example of these concerned businesses.

Long touted as a socially responsible business, Ben and Jerry's buys its ice cream ingredients from small U.S. farmers and family businesses. But Ben and Jerry's also buys ingredients from extractive rainforest reserves.

The idea to put ingredients from the rainforest into an ice cream sold in the U.S. originated in 1988 at a rainforest benefit party. There, Ben Cohen, co-founder of Ben and Jerry's, happened to meet Jason Clay, the director of Cultural Survival, a non-profit organization dedicated to preserving indigenous cultures in their natural environments.

Clay recently had returned from the Amazon. He had with him samples of exotic nuts harvested by Indians from Brazil's extractive reserves. Clay's intention was to convince American businessmen like Cohen that by marketing the nuts they could not only profit but also help protect rainforests and indigenous cultures.

Cohen was enthusiastic about the idea. He took it upon himself to devise a recipe using the nuts and subsequently ordered 15 tons of nuts for his company. Shortly, thereafter, his best-known product was an ice cream made with the nuts, called Rainforest Crunch.

Through a separate company, Community Products Inc., Cohen also markets two other products that contain the nuts, Rainforest Crunch candy bars and Rainforest Crunch popcorn. In all, Cohen's enterprises purchase 150 tons of nuts each year from the rainforest.

The Body Shop

Another business that buys products from extractive reserves is The Body Shop, a British cosmetic company. The Body Shop uses a number of ingredients from the rainforest in its skin softeners, bath beads, lip balms, conditioners, and body lotions. The rainforest ingredients in these products include andiroba oil, piquia oil, babassu oil, Brazilian nut oil, naja oil, murumuru palm tree kernels, urucum seeds, honey, and beeswax.

Anita Roddick, The Body Shop's founder and managing director, believes that by creating a market for these items her company is creating an economic incentive to save rainforests.

Like Cohen, Roddick also uses her business to support indigenous cultures. She does this by purchasing the natural ingredients that her company uses from Indians. These Indians include the Kayapo of Brazil, the Nanhu of Mexico, the Santa Ana Pueblo of New Mexico, and native communities in the Solomon Islands, Nepal, Tanzania, and Zambia.

In 1990 Roddick's support of sustainable development was recognized by the United Nations' "Global 500" environmental award.

Whole Foods Market

Sustainably producing products from the rainforest is only part of the solution. The ultimate goal of sustainable development is to produce all the products that our society needs in an earth-friendly manner.

Such a goal might seem impossible, but a company called Whole Foods Market is pioneering this effort. Whole Foods is the largest chain of natural foods grocery stores in the United States. On average, each of its stores markets 12,000 food and non-food products produced with concern for the environment. Among the products sold at Whole Foods is Ben & Jerry's Rainforest Crunch ice cream.

Public Support

Public support for these types of businesses is tremendous. From 1983 to 1992, Ben & Jerry's yearly sales rose from $1.8 million to $132 million. In the last ten years, The Body Shop's annual sales increased from £2.1 million to £168.3 million. Since opening its first store in 1980, Whole Foods has grown to include some 43 stores in ten U.S. states, selling over $400 million worth of natural and organic products annually.

Sustainable businesses, however, represent only a fraction of the global economy. But by buying more of the products that we use daily from these businesses, we, as consumers, can help these businesses save rainforests.

Ben Cohen explains, "Businesses have created most of our social and environmental problems. If businesses were instead trying to solve these problems, they would be solved in short order."

Government Regulations

Governments can help save rainforests through regulations on commerce that protect the environment. These "earth-friendly" regulations would both require and encourage businesses to pay for using natural resources.

Paying for Resources

One way governments can require businesses to pay for using natural resources is through user fees or permit fees. Regulations like these are based on an economic concept called "natural resource accounting."

The premise behind this concept is that natural resources, such as clean air, pure water, and virgin forests, have a monetary value. Consequently, the price of a permit to destroy these natural resources should equal the cost of replacing them.

For example, lumber companies (that do not grow their own trees) should be required to buy a user fee equal to the cost of planting and raising to maturity the trees they harvest.

Similarly, companies that create water pollution should pay for a pollution permit, whose cost would equal the cost of cleaning up their pollution.

Incentives for Conservation

In addition to penalties, governments can encourage businesses to operate without damaging the environment through financial rewards or incentives. These rewards could take many forms, including government-backed loans or direct subsidies. But the most comprehensive way for governments to encourage sustainable development is through tax policy.

Preferential tax treatment needs to be given to businesses that make products with few natural resources. For example, recycled paper producers should be taxed less than virgin paper producers. Currently in the U.S., recycled paper is slightly more expensive than non-recycled paper. (Printing this book on recycled paper cost an additional 8 percent.) With the appropriate tax policy, recycled paper could be made to cost less than virgin paper.

Preferential tax treatment also could make lumber from tree farms less expensive than lumber from old growth forests. Such policies could even encourage firms to develop paper and wood substitutes. In fact, any product deemed to be environmentally friendly—alternative fuels, electric cars—should be subject to a minimum of taxation.

Incentives for Green Investing

Another tax policy that can protect the environment is to give tax credits and tax deductions to people who invest in earth-friendly businesses. Profits from such investments could be made subject to low capital gains tax rates, as well.

Who Pays the Tax?

These tax incentives and disincentives are basically consumption taxes; individuals and businesses that consume products made with few natural resources would end up paying fewer taxes.

This does not mean that our society overall must pay more or less in taxes. The only change would be in who pays the tax.

For example, a tax policy based on sustainable development might decrease personal income taxes while increasing the tax on gasoline; theoretically, society as a whole would pay the same amount in tax but would tend to conserve energy.

Levying taxes on industries that pollute is another policy that encourages sustainable development. Revenues from these taxes would be earmarked to clean up pollution. This is known as the "polluter pays principle."

In favor of the "polluter pays principle" is the World Resources Institute, an organization of scientists and policy experts supported by the United Nations and private foundations. This institute recommends raising taxes on social ills, such as pollution, while lowering taxes by an equal amount on social virtues, such as work.

Why Not Ban Products from the Rainforest?

Government regulations that require businesses to pay for natural resources and pollution are controversial ideas. Even some environmentalists disagree with these ideas.

These people feel that products harmful to the environment should be banned outright instead of regulated and taxed.

In some cases, products should be banned, i.e., pesticides that are overly toxic (like DDT). But if products from the rainforest are banned, deforestation could increase.

This is because industries in tropical countries, such as timber, agriculture, and cattle production, make up a large portion of tropical economies. Banning the products of these industries would leave many people unemployed, thereby increasing poverty and the number of subsistence farmers who clear rainforests to survive.

This is why sustainable development needs to be integrated into the economic system of each nation.

International Trade Regulations

Rainforests (and the environment) can be protected through new regulations on international trade.

These regulations need to be coordinated among all nations and would penalize through trade restrictions those nations whose industries produce goods non-sustainably.

Trade Agreements

Mechanisms for enforcing this international effort already exist—the trade agreements that currently govern international trade. The focus of most trade agreements, however, needs to change.

Trade agreements, i.e., trade treaties, typically focus on the short-term; their goals are to increase corporate profits, not promote long-term development.

For example, the North American Free Trade Agreement (NAFTA), which recently lowered trade barriers among Canada, the United States, and Mexico, is allowing many American businesses to relocate to Mexico to avoid costly environmental regulations in the U.S. Eventually, these businesses will damage Mexico's environment, while their products, for the most part, will be consumed by people in the United States and Canada.

In the long term, such a trade relationship can only worsen living conditions in Mexico and damage the Mexican economy. This is why many U.S. environmental organizations, as well as many Mexicans, are opposed to the changes brought about by NAFTA.

To promote sustainable development, NAFTA needs to be amended so that free trade is possible only for goods whose production does not damage the environment in any of the three countries.

NAFTA is just one of many trade agreements that can help promote sustainable development. Other such agreements include the General Agreement on Tariffs and Trade (GATT), the Asia-Pacific Economic Cooperation Agreement (APEC), and the agreements being discussed under the Enterprise for the Americas Initiative (EAI).

Of these agreements, GATT is the most comprehensive. Signed by more than 120 nations, GATT is a treaty that governs about 90 percent of all internationally traded goods and is updated periodically. (Future GATT agreements will be the responsibility of the newly created World Trade Organization)

APEC, which is under negotiation, is an attempt to create free-trade among 18 countries bordering the Pacific Ocean, including the United States.

The EAI, which is also under negotiation, is an effort by the United States to expand NAFTA to include all the countries of North and South America.

Trade agreements, such as APEC, GATT, NAFTA, and those being considered under EAI, all should place restrictions on goods produced at the expense of rainforests and the environment.

Won't This Hurt Us Economically?

All of these international agreements are part of a larger effort by international businesses to increase free-trade around the world. Conventional wisdom holds that doing this will make everyone better off.

Herman Daly, an economist for the World Bank, disagrees. He warns, "Further growth beyond the present scale is overwhelmingly likely to increase costs more rapidly than it increases benefits, thus ushering in a new era of 'uneconomic growth' that impoverishes rather than enriches."

What Daly is saying is that increasing trade in goods produced from resources that are not replaced will make us worse off.

The reason for this paradoxical situation is that most countries do not take into account the value of natural resources when they measure their economies. Their gross national product (GNP), a measure of economic vitality, only measures what they produce.

For example, if a country's forests are destroyed to produce lumber, that country's GNP will increase when that lumber is sold—but GNP will not record the loss in forest resources.

Robert Repetto, a policy expert for the World Resources Institute notes, "A country could exhaust its mineral resources, cut down its forests, erode its soils, pollute its aquifers, and hunt its wildlife and fisheries to extinction, but measured income would not be affected."

Recalculating GNP

Calculations of a country's GNP need to reflect losses in natural resources, including the loss of rainforests. This is, in essence, natural resource accounting on a national level.

Some countries—Australia, Canada, France, the Netherlands, and Norway—already are implementing a national accounting of their natural resources. In addition, the United Nations Statistical Commission is creating guidelines for countries to measure the costs of replacing natural resources. Various international organizations, such as the World Resources Institute, are doing the same.

When our existing economies become sustainable, then we can proceed toward global free-trade. Otherwise, increases in trade will only diminish our planet's stock of natural resources and accelerate the destruction of our rainforests.

What We Can Do

Domestic trade regulations and international trade treaties that protect the environment are powerful agents of change. As individuals, we can vote for politicians and political parties that support these ideas or become active in one of the many environmental organizations that lobby for such policies. A list of these organizations is included in the appendix of this book.

RETHINKING INTERGOVERNMENTAL LENDING

Changing the focus of the World Bank and the IMF, can help save rainforests. These international lending institutions, which were originally created to rebuild the economies damaged by World War II, must now rebuild a world damaged ecologically and largely impoverished.

Past Policies

In the recent past the World Bank and the IMF have encouraged many less-developed countries to produce and export natural resources, such as timber, cattle, and agricultural products, to repay debts owed to foreign commercial banks. On average, less-developed countries still spend about 30 percent of the money they earn from exports just to repay interest on their debts. (In 1986, Mexico and Argentina spent over 50 percent of their export earnings to service foreign debts.)

Largely because of these policies, many underdeveloped countries have become economic and environmental disasters. They have sold off their natural resources and tried to save money by cutting social programs.

Within these countries are not only high rates of tropical deforestation but also some of the poorest living conditions in the world. It is estimated that in less-developed countries over 700 million people are malnourished and more than one billion people lack proper sanitation and safe water.

Selling the natural resource assets of these people to repay foreign debts is a policy that should be abandoned.

Appropriate Policies for the IMF and World Bank

The mandate of the IMF and World Bank has never been to rescue commercial banks that make bad loans. The purpose of these institutions is, rather, to improve human conditions. This includes environmental conditions.

Appropriate lending policies should help tropical countries develop self-sustaining economies that are not oriented around the export of natural resources.

The IMF and World Bank need to help troubled countries produce food for local markets, create diversified economies, and build industries that produce goods for domestic use. Also, less-developed countries must be allowed to protect their industries by restricting the operations of foreign companies within their borders.

In addition, appropriate policies would include investments in people, i.e., support for small businesses and technological improvements, financial assistance for education and job-training, and more spending on health care.

In the long run, these types of policies will do more to alleviate poverty than do the current policies of encouraging impoverished people to migrate into rainforests.

All of these policies will help build self-sustaining countries.

World Bank Environmentalism

To some degree, the World Bank is taking steps to support sustainable development and protect the environment.

In 1989 the World Bank began making what it called "an environmental assessment" on all projects that could have a negative impact on the environment.

In 1994, $748 million in loans were approved for projects to improve the environments of underdeveloped countries.

The World Bank also changed its lending policy toward commercial logging in tropical forests. It did this after several conservation groups (including Friends of the Earth and the World Wide Fund for Nature) started a letter-writing campaign to protest loans for logging rainforests in

U.S. Policy on Debt Forgiveness
Until recently, the United States only supported World Bank efforts to refinance, rather than forgive, the debts of financially troubled countries. But in 1989, U.S. Treasury Secretary Nicholas Brady announced that the United States would begin to work with the World Bank to arrange debt forgiveness for impoverished countries.

This new policy was known as the Brady Plan. As part of the Brady plan, about half of Mexico's foreign debt was restructured, forgiving about $12 billion worth of future payments.

Where Mexico was concerned, the U.S. felt it could benefit from a stable, southern neighbor. A stronger Mexican economy would result in more demand for U.S. goods and fewer illegal immigrants from Mexico.

Since the U.S. announced its new policy, debt forgiveness has become a viable, although under-used, option for the World Bank. To date, the World Bank has made debt forgiveness arrangements for Argentina, Mexico, Jordan, Mozambique, Guyana, Uganda, Bolivia, Costa Rica, the Philippines, Venezuela, Uruguay, and Niger.

The Brady Plan, however, has fallen short of its goal, which was to reduce the debts of 38 countries by 20 percent over three years. A renewed effort at debt forgiveness is needed if we are to decrease the pressure on less-developed countries to sell off their forest resources.

Debt-For-Nature Swaps
On a smaller scale, various environmental organizations are helping alleviate the debt burdens of some countries, while at the same time creating rainforest preserves.

These organizations, often collaborating with each other, buy from international banks a portion of the debt of an impoverished country; they then "forgive" the debt when the country sets aside a nature preserve. This exchange is known as a debt-for-nature swap.

Typically, international banks sell the debts of troubled countries at steep discounts because they know that full repayment is unlikely anyway. These discounts allow environmental organizations to create nature preserves in foreign countries without spending large amounts of money.

The first "debt-for-nature swap" occurred in 1987, when Conservation International purchased $650,000 worth of Bolivian debt for 15 cents on the dollar. This debt was cancelled when Bolivia set aside 2.7 million acres in conservation areas.

In 1991 the Rainforest Alliance also made such an arrangement. This organization bought Costa Rican debt from the Central American Bank for Economic Integration. The Rainforest Alliance then exchanged the debt for Costa Rican land and created the International Children's Rainforest.

In total, debt-for-nature swaps have resulted in the forgiveness of $98.4 million in debt but have cost conservation organizations only $16.7 million.

Because of debt-for-nature swaps, wildlife preserves have been established in Bolivia, Costa Rica, the Dominican Republic, Ecuador, Ghana, Poland, Jamaica, Madagascar, Mexico, Nigeria, the Philippines, and Zambia.

Pioneering this effort are Conservation International, the Conservation Trust of Puerto Rico, Debt for Development Coalition, the National Park Foundation of Costa Rica, Monteverde Conservation League, Missouri Botanical Gardens, the Rainforest Alliance, the Smithsonian Institute, the Nature Conservancy, and the World Wildlife Fund.

The Sustainable Development Solution

The destruction of rainforests can be stopped. But it will require a change in our thinking. We can no longer view less-developed countries as a source of raw materials for the products we, in the developed world, want to buy. The World Bank and IMF need to help these countries build self-sustaining economies that do not damage the environment.

We need to realize that wiping out rainforests to pay off banks is a poor exchange. Through debt forgiveness and debt-for-nature swaps, we can reduce the debts of underdeveloped tropical countries and remove the pressures on them to sell their natural resources.

Our economic systems need to recognize the value of a healthy environment. Government regulations can be used to penalize businesses that damage the environment and reward those that produce goods sustainably.

Even our environmental efforts must change. Through ecotourism and extractive reserves, we can protect more natural areas by producing more income from these areas.

Our planet's resources are limited, and we are devouring them at an increasing rate. This can change, however, if we simply replace resources as we consume them. This is the sustainable development solution.

LEADERS IN THE COMING ECOLOGICAL ERA

At first I thought I was fighting to save rubber trees, then I thought I was fighting to save the Amazon rainforest. Now, I realize I am fighting for humanity.

—Chico Mendes

Today, we must incorporate into our modern economy a philosophy of sustainability. We need to live off the interest of our planet's assets, not the principal. With a growing population and dwindling natural resources, our future—and our children's future—depend on these things.

Creating such a world is the essence of sustainable development, the future of environmentalism, and the challenge of our generation.

POLITICAL LEADERSHIP

Recently, a political organization rose to this challenge and proposed that sustainable development be implemented worldwide.

This organization is recommending that all nations accept certain principles of environmental protection and begin work to conserve forests. This organization has asked that all nations sign treaties to protect endangered species, preserve natural habitats, and reduce carbon dioxide emissions. This organization wants all nations to accept legal responsibility for their private companies that damage the environments of foreign countries.

Moreover, this political organization aspires to create a worldwide economy based on the sustainable use of natural resources in the 21st century.

This political organization is the United Nations, and its proposals were put forth at the United Nations Conference on Environment and Development, better known as the Earth Summit.

The Earth Summit

Held in Rio de Janeiro, Brazil, in 1992, the Earth Summit was a meeting of more than 100 presidents, princes, prime ministers, and other national leaders from around the world. It was the largest international conference on environmental issues in history.

The Earth Summit's plan for a sustainable world is called *Agenda 21*. It is an 800-page document available in book stores everywhere.

THE ROLE OF THE NON-GOVERNMENTAL ORGANIZATION

Thousands of non-governmental organizations (NGOs) are working to teach people about sustainable development. These NGOs, whose influence transcends national boundaries, include a variety of organizations, such as the Cousteau Society, Sierra Club, National Audubon Society, and Greenpeace.

At the same time that government leaders were meeting at the Earth Summit, more than 7,000 NGOs from 165 different countries held a conference for the general public in Rio's Flamingo Park. Called the Global Forum, this conference was a series of events, seminars, and lectures that covered the same issues the United Nations addressed across town. More than 200,000 people attended the Global Forum.

Leadership in the Business World

Many businesses are taking the lead to create a sustainable world. Corporate leaders include Federal Express, which is testing in over 100 of its delivery vans the use of alternative fuels that create less pollution than gasoline.

Also, the McDonald's Corporation is working with the Environmental Defense Fund to recycle trash and reduce the quantity of fast-food packaging at McDonald's restaurants. McDonald's now boasts that it has purchased over $1 billion worth of recycled paper.

Investment professionals at the Co-operative Bank of Manchester, England, are a part of this leadership. This bank will not lend money to companies involved in illegal activities, arms manufacturing, trade with oppressive regimes, animal testing, or any environmentally destructive practice.

Mutual funds are even being created to support earth-friendly companies. The New Alternative Mutual Fund, for example, invests in solar and alternative energy companies. The Freedom Environmental Mutual Fund invests in pollution-control and waste-management businesses.

Grass-Roots Support

Around the world, grass-roots efforts are rising to the challenge of sustainability.

In China, for example, local communities have planted a forest 250 to 1,000 miles wide and 4,300 miles long to stop an encroaching desert. Called the San Bei forest belt, this reforestation effort eventually will cover 1.4 million square miles with trees.

In the former Soviet Union, local environmentalists have blocked government plans to pipe waste into the Irkutsk River. At Krasnoyarsk, Russia, 60,000 residents signed a petition to block construction of a nuclear facility they feared would become another Chernobyl. Currently, Soviet citizens are protesting the dumping of industrial waste by a state-run paper factory into Lake Baikal, the largest fresh-water lake in the world.

In Europe, environmentalists have created their own political parties and are winning elections. Green Party candidates have been elected to national parliamentary offices in Switzerland, Belgium, Finland, Portugal, Germany, Luxembourg, Austria, Italy, Sweden, the Netherlands, and Greece.

In the United States, voters are supporting the green movement, as well. In the 1992 presidential election, Bill Clinton received widespread support for choosing the senate leader on environmental issues, Al Gore, to be his vice presidential running mate. Before becoming vice president, Senator Gore was not only a supporter of environmental legislation but also the author of a best-selling environmental book, *Earth in the Balance*.

All over America people are choosing to protect their environment. In Portland, Oregon, citizens forced their city council to protect 100-year-old trees, called "Heritage Trees." In Seattle, Washington, residents created a curbside recycling program that has become the model for the rest of the nation. In Austin, Texas, three out of seven current city council members were elected on environmental platforms.

In these cities and thousands of others throughout the world, environmentalism is growing.

INDIVIDUALS AS LEADERS

Countless individuals are taking up the challenge to solve our environmental problems. These people are everywhere.

Frances Lynn Carpenter, an ecology professor from the University of California, Irvine, invested her life savings to buy a deforested Costa Rican cattle ranch and is there replanting a rainforest.

Wangari Maathai of Kenya organized school children and women's groups to plant more than 10 million trees in 30 African countries. Her effort, called the "Green Belt Movement," also teaches African women about nutrition and family planning.

Professor Sandra Neil Kinghorn founded a sustainable development school in Central American. Kinghorn's school teaches students from around the world about rainforests and the environment.

Max Nofziger works as an Austin, Texas, city councilman to protect a renowned natural swimming pool, Barton Springs, from the pollution of real-estate development. A former street-corner flower salesman, Nofziger ran unsuccessfully for city council four times on environmental issues. On his fifth attempt he was elected and has since been re-elected twice more. Nofziger and Barton Springs have now come to symbolize environmental protection in central Texas.

LEADERS IN THE COMING ECOLOGICAL ERA 113

Michael Werikhe of Kenya is teaching people around the world about the need to protect the endangered Black Rhino. Formerly an employee for the Kenyan Game Department, Werikhe became so frustrated by the relentless poaching of Black Rhinos that he quit his job and walked 300 miles through the Kenyan countryside, persuading villagers not to harm rhinos. Werikhe then continued his crusade by walking through Tanzania, Uganda, Italy, Switzerland, Germany, the Netherlands, England, and now the United States—over 4,000 miles on foot—delivering his message to millions of people.

Nine-year-old Roland Ticnsuu of Sweden helped create a rainforest preserve in Costa Rica. Tiensuu persuaded his elementary school class to raise money for the Monteverde Conservation League, a nonprofit organization that buys and preserves rainforests. Tiensuu's class raised $300 by selling books and paintings, which they created, and tickets to concerts, where they preformed.

Tiensuu's teacher, Eha Kern, subsequently founded a non-profit organization of her own, called Barnens Regnskog (Children's Rainforest), which has grown to include some 6,000 schools throughout Europe. Children from these schools have raised more than $2 million, and purchased nearly 26 square miles of rainforest in Costa Rica.

These individual leaders also include Chico Mendes. An impoverished, Brazilian, rubber-tree worker, Mendes organized a union of his co-workers to protest the wholesale clearing of rainforests by logging and cattle companies in the Amazon. His union persuaded the Brazilian government to establish millions of acres of extractive reserves, including one in the Seringal Cachoeira rainforest where Mendes was born and raised.

Mendes received many death threats from those opposed to his efforts. Several times he was imprisoned and beaten by local police. In an interview, Mendes was asked about threats on his life. He remarked philosophically, "If an angel came down from the sky and could guarantee that my death could strengthen this fight, it would be a fair exchange."

On the night of December 22, 1988, while celebrating the Christmas holidays with his family, Mendes was assassinated.

Millions of other people are working to protect the environment: people who are ordinary citizens in everyday communities; people who organize recycling efforts, plant trees in their neighborhoods, and pick up litter.

All of these people are leaders in a worldwide cultural movement that is creating a sustainable world. These people are helping us realize the importance of the environment. They give us a vision of a better world, and they evoke in us our capacity to care.

TO ORDER ADDITIONAL COPIES

If you would like additional copies of this book, send $9.95 plus $2 shipping and handling per copy to:
> The Better Planet Press
> P.O. Box 160146
> Austin, Texas 78716

APPENDIX

ORGANIZATIONS WORKING TO PRESERVE RAINFORESTS

U.S. Organizations

African Wildlife Foundation, 1717 Massachusetts Ave. NW, Washington, DC 20036.

American Forests, P.O. Box 2000, Washington, DC 20013.

Center For Environmental Information, 46 Prince st., Rochester, NY 14607.

Conservation International, 1015 18th st. NW Ste. 1000, Washington, DC 20036.

The Cousteau Society, Inc., 870 Greenbrier Circle, Ste. 402, Chesapeake, VA 23320.

Cultural Survival, 11 Divinity Ave., Cambridge, MA 02138.

Defenders Of Wildlife, 1244 19th st. NW, Washington, DC 20036.

Ducks Unlimited, Inc., One Waterfowl Way, Memphis, TN 38120.

Earth Island Institute, 300 Broadway, Ste. 28, San Fancisco, CA 94133.

Earthwatch, P.O. Box 403N, 680 Mt. Auburn st., Watertown, MA 02172.

Environmental Action, 6930 Carroll Ave., Ste. 600, Takoma Park, MD 20912.

Environmental Data Research Institute, Inc. 797 Elmwood Ave., Rochester, NY 14620.

APPENDIX

Environmental Defense Fund, 257 Park Ave. S., New York, NY 10010.
The Environmental Exchange, 1930 18th st., NW, Ste. 24, Washington, DC 20009.
Environmental Support Center, Inc., 1875 Connecticut Ave. NW, Ste. 340, Washington, DC 20009.
Friends Of The Earth, 218 D st., SE, Washington, DC 20003.
Global Tomorrow Coalition, 1325 G st., NW, Ste. 915, Washington, DC 20005.
Great Old Broads for Wilderness, P.O. Box 520307, Salt Lake City, UT 84152.
Green Peace USA, 1436 U St. NW, Washington, DC 20009.
International Alliance for Sustainable Agriculture, 1701 University Ave. SE, Minneapolis, MN 55414.
National Arbor Day Foundation, 211 N. 12th Street, Suite 501, Lincoln, NE 68508.
National Audubon Society, 700 Broadway, NY, NY 10003.
National Audubon Society Expedition Institute, P.O. Box 365, Belfast, ME 04915.
National Wildlife Federation, 1400 16th st. NW, Washington, DC 20036.
Native Forest Council, P.O. Box 2171, Eugene, OR 97402.
National Resources Defense Council, 40 W. 20th st., New York, NY 10011.
The Nature Conservancy, 1815 N. Lynn st., Arlington, VA 22209.
Pele Defense Fund, P.O. Box 404, Volcano, HI 96875.

Population-Environmental Balance, 1325 G st. NW, Ste. 1003, Washington, DC 20005.

Rainforest Action Network, 450 Sansome st., Ste. 700, San Francisco, CA 94111.

Rainforest Alliance, 270 Lafayette st., Ste. 512, New York, NY 10012.

Sierra Club, 730 Polk st., San Francisco, CA 94109.

Sierra Club Legal Defense Fund, 180 Montgomery st., Ste. 1400, San Francisco, CA 94104.

Southwest Network For Environmental And Economic Justice, P.O. Box 7399, Albuquerque, NM 87194.

Student Conservation Association, Inc., P.O. Box 550, Charlestown, NH 03603.

Thorne Ecological Institute, 5398 Manhatten Cir., Boulder, CO 80303.

20/20 Vision, 30 Cottage st., Amherst, MA 01002.

Union Of Concerned Scientists, 26 Church st., Cambridge, MA 02238.

The Wilderness Society, 900 17th st. NW, Washington, DC 20006.

Wildlife Conservation International, c/o New York Zoological Society, Bronx, NY 10460.

The Wildlife Society, 5410 Grosvenor Ln., Bethesda, MD 20814.

Women's Environment And Development Organization, 845 Third Ave., 15th floor, New York, NY 10022.

World Resources Institute, 1709 New York Ave., NW, Ste. 700, Washington, DC 20006.

World Society For The Protection Of Animals, P.O. Box 190, Boston, MA 02130.

World Wildlife Fund, 1250 24th st., NW, Ste. 400, Washington, DC 20037.

Zero Population Growth, 1400 16th st., NW, Ste. 320, Washington, DC 20036.

Foreign Organizations

Campa a Amazonia: Por La Vida, P.O. Box 246C, Quito, Ecuador.

Environmental Foundation Ltd., 6 Boyd Place, Colombo 3, Sri Lanka.

Friends Of The Earth/U.K., 2628 Underwood st., London N17JU, United Kingdom.

Haribon Foundation, Suite 306, Sunrise Condo, Ortegas Ave., San Juan, Metro Manilla, Philippines.

Indonesian Environmental Forum, J1, Penjernihan I, Kompl. Kenangan 15, Penjompongan, Jakarta 10210, Indonesia.

International Union For The Conservation Of Nature And Natural Resources, Avenue Mont Blanc, 1196 Gland, Switzerland.

Research Institute For Natural Resource Policy, 105 Rajpur Rd., Dehra Dun, Uttar Pradesh 248001, India.

World Rainforest Movement, 87 Contonment Road, 10250 Penang, Malaysia.

World Wildlife Fund/U.K., Panda House, Godalming, Surrey, GU7 1XR, United Kingdom.

REFERENCES (IN ORDER OF APPEARANCE)

PREFACE

Edward.O. Wilson, Ed., *Biodiversity* (Washington D.C.: National Academy Press, 1988).

Edward O. Wilson, *The Diversity of Life* (New York: W.W. Norton & Company, Inc.,1992) (Beebe quote on page 140).

John A. Hoyt, *Animals in Peril: How "Sustainable Use is Whiping Out the World's Wildlife* (Garden City Park, NY: Avery Publishing Group, 1994).

Stephen R. Kellert and Edward O. Wilson, eds., *The Biophilia Hypothesis* (Washington D.C.: Island Press, 1993).

Lester R. Brown, Nicholas Lessen, and Hal Kane, *Vital Signs 1995* (New York: W.W. Norton & Company, Inc., 1995).

Julian Baum, "Refusing the Bait: Fishing Industry Set to Ignore Driftnet Ban," Far Eastern Economic Review, January 23, 1993.

George Wehrfritz, "Gone Fishing: Rogue Trawlers may be Dodging Official Driftnet Ban," Far Eastern Economic Review, Febuary 25, 1993.

Melana Zylam, "Deep Trouble: Russian Nuclear Waste Dumped in Sea of Japan," Far Eastern Economic Review, March 18, 1993.

Rainforest Alliance, "There's Still Time," a Rainforest Alliance pamphlet, The Rainforest Alliance, 270 Lafayette Street, suite 512, New York, NY 10012.

REFERENCES

CHAPTER 1
  Wilson, *The Diversity of Life.*

CHAPTER 2
  Sting, "Primal Sting (Sting's South American Tour)," Vogue, June 1988 (Tacuma quote on page 304).
  Wilson, *The Diversity of Life.*
  Norman Myers, "Tropical Forests: Present Status and Future Outlook," Climatic Change, September 1991.
  Alan Grainger, "Rates of Deforestation in the Humid Tropics: estimates and measurements," The Geographical Journal, March 1993.
  Brown, Lessen, and Kane.
  Norman Myers, *The Gaia Atlas of Future Worlds* (London: Robertson McCarta Limited, 1990).
  Geoffrey Lean, Polly Ghazi, and Mary Lean, Forest supplement, produced by World Wide Fund for Nature, UK, and the Observer, WWF/UK, Panda House, Godalming, Surrey, GU7 1XR, United Kingdom.
  Scott Lewis, *The Rainforest Book* (Venice CA: Living Planet Press, 1990).
  David Linders and Elizabeth Lawrence, "A Growing Concern," a Rainforest Alliance pamphlet, Rainforest Alliance, 270 Lafayette Street, suite 512, New York, NY 10012.
  Naomi Rosenblatt, *Rainforests For Beginners* (New York: Writers and Readers Publishing, Inc., 1992).
  Cultural Survival, "Cultural Survival Catalog 1993," Cultural Survival, 215 First Street, Cambridge, MA 02142.

CHAPTER 3
  Julian Burger, *The Gaia Atlas of First Peoples* (New York: Anchor Books, Doubleday, 1990) (Paiakan quote on page 32).

Myers, "Tropical Forests."
Brown, Lessen, and Kane.
Lean, Ghazi, and Lean.
Lewis.
Malcolm Gillis and Robert Repetto, eds., *Public Policies and the Misuse of Forest Resources* (New York: Cambridge University Press, 1988).
International Tropical Timber Organization (ITTO), The division of Economic Information and Market Intelligence, Annual Review and Assessment of the World Tropical Timber Situation 1992 (ITTO: Yokohama, Japan).
Hand Book of Economic Statistics 1991 (Directorate of Intelligence CIA, Aug 1, 1991).
Stan Sesser, "A Reporter at Large: Logging the Rainforest," The New Yorker, May 27, 1991.
Hal Kane with Linda Starke, *Time for Change* (Washington D.C.: Island Press, 1992).
Jose Serra-Vega, "Andean Settlers Rush for Amazonia," Earthwatch, 3rd Quarter 1990, no 39.
"Across a Gap in Darien," The Economist, November 21, 1992.

CHAPTER 4

Linda J. Seligmann, "The Burden of Visions Amidst Reform: Peasant Relations to Law in the Peruvian Andes," American Ethnologist, Febuary 1993 (Borah quote on page 25).
Paul R. Krugman, "Debt Relief is Cheap," Foreign Policy, Fall 1990, n. 80.
Lewis W. Snider, "The Political Performance of Third World Governments and the Debt Crisis," American Political Science Review, December 1990.

REFERENCES

Miles Smith-Morris, ed., *The Economist, Book of Vital World Statistics* (New York: Times Books, 1990).

Pennwell Publishing Co., *International Petroleum Encyclopedia 1992*, vol 25 (Tulsa, OK: Pennwell Publishing Co., 1992).

United Nations Department of Economic and Social Information and Policy Analysis, *World Economic Survey 1993* (New York: United Nations, 1993).

John Walton and Charles Ragin, "Global and National Sources of Political Protest: Third World Responses to the Debt Crisis," American Sociological Review, 1990, vol 55.

Z. Cernohous, "International Banks and Funds," *Collier's Encyclopedia*.

Lisa Beyer, Steven Gutkin/Caracas and John Moody/San Jose, "Crackdown in Caracas," Time, March 13, 1989.

"The Morning After Oil Binge," The Economist, March 4, 1989.

The World Bank, *The World Bank Annual Report 1994*, The World Bank, 1818 H Street, N.W., Washington, D.C. 20433.

Burger.

Gillis and Repetto.

Myers, "Tropical Forests."

Brown, Lessen, and Kane.

Stephan Schmidheiny, *Changing Course* (Cambridge, Massachusetts: The MIT Press, 1992).

The Economist Books Ltd, *Book of Vital World Statistics* (Times Books/Ramdom House, 1990).

United Nations.

International Monetary Fund, *1994 International Monetary Fund Annual Report*, IMF, 700 19th st., N.W., Washington, D.C. 20431.

Lewis.

CHAPTER 5

Russell Wild ed., *The Earth Care Annual 1992* (Emmaus, Penn.: Rodal Press, Inc., 1992) (Hair quote on page ix).

Photos and caption provided by NASA, Houston, Texas

CHAPTER 6

Joseph R. Conlin, *Morrow Book of Quotations in American History* (New York: William Morrow and Company, Inc., 1984) (Tecumseh quote on page 281).

President Andrew Jackson, *Messages of Gen. Andrew Jackson* (Boston: Otis Broaders and Company, 1837, also Concord, N.H.: John F. Brown, and William White, 1837).

Dee Brown, *Bury My Heart at Wounded Knee* (New York: Holt, Rinehart, and Winston, 1970).

Jonathan H. Adler, "Popular Front: The Rebirth of American Forests," Policy Review, Spring 1993.

John Elk vs Charles Wilkins (U. S. Supreme Court, Nov. 3, 1884).

"Andrew Jackson," *World Book Encyclopedia*, 1994 ed.

Alvin M. Josephy Jr., ed., *The American Heritage Book of Indians* (New York: American Heritage Publishing Co. Inc., 1961).

Gillis and Repetto.

Jay D. Hair, "Habitat Destruction Dooms People Too," International Wildlife, March/April 1991.

Lean, Ghazi, and Lean.

Stan Sesser, "A Reporter at Large: Logging the Rainforest," The New Yorker, May 27, 1991 (Sesser gives ITTO estimate of Sarawak's forest depletion on page 47)

REFERENCES

(Prince Charles quote on page 43) (Sesser quotes on page 66 and 61 respectively).

ITTO.

Napoleon A. Chagnon, "Life Histories, Blood Revenge, and Warfare in a Tribal Population," Science, February 26, 1988.

William Booth, "Warfare Over Yanomamo Indians," Science, March 3, 1989 (Quote about Yanomami on page 1138).

Nancy Stoetzer, "In Search of a Refuge," Buzzworm: The Environmental Journal, May/June 1991 (Carias quote on page 23).

Bob Levin and Richard House, "An Imperilled People," Maclean's, June 29, 1987.

"The Indians and the Forest," Commonweal, May 20, 1988.

Andrea Dorfman, "Assault in the Amazon," Time, Nov. 5, 1990.

Ann Gibbons, "Yanomami people threatened," Science, June 21, 1991.

Marlise Simons, "The Amazon's Savy Indians," New York Times Magazine, February 26, 1989.

Alexander Cockburn, "Beat the Devil," The Nation, November 7, 1988.

Carl Zimmer, "Tech in the Jungle," Discover, August 1990.

Norman Myers, "Kuna Indians: Building a Bright Fruture," International Wildlife, July/August 1987.

The Futurist, "Biosphere Reserves," May-June 1992.

"South America: Saving the Yanomami," UN Chronicle, June 1993.

CHAPTER 7

Schmidheiny (Einstein quote on page 82).

Hal Kane with Linda Stark, *Time for Change* (Washington D.C.: Island Press, 1992) (Daly quote on page 60) (Repetto quote on page 64) (Our Common Future quote on page 11).

Lean, Ghazi, and Lean (World Bank policy statement on page 33).

Debby Crouse, "Up Close with Gorillas," International Wildlife, Nov/Dec 1988.

Wilson, *The Diversity of Life.*

Ben & Jerry's Homemade Inc., Ben & Jerry's information packet, received January 25, 1994, Ben & Jerry's, route 100, P.O. Box 240, Waterbury, Vermont 05676.

Community Products Inc., Statements on Rainforest Crunch Popcorn package, a product of Community Products Inc.

Pamphlet from Community Products Inc.; Community Products Inc., RD # 2, Box 1950, Montpelier, Vermont 05602.

Paul Hawken, "A Declaration of Sustainability," Utne Reader, Sep/Oct 1993 (Ben Cohen quote on page 72).

The Body Shop, This is the Body Shop pamplet, spring 1994, The Body Shop, 45 Horsehill Road, Cedar Knolls, NJ 07927.

Standard and Poor's Equity Research Department, *Standard and Poor's Over the Counter and Regional Stock Exchange Reports*, vol 59, n 134, Sec 63, Nov 24, 1993 (New York: McGraw-Hill Inc., 1993).

Standard and Poor's, *Standard Corporations Descriptions* (New York: McGraw-Hill Inc., January 13, 1994).

REFERENCES

Whole Foods Market, 1994 Annual Stakeholders Report, Whole Foods Market, Inc., 1705 Capital of Texas Highway South, Suite 400, Austin, TX 78746.

Telephone conversations with Whole Foods, August 29, 1995.

World Resources Institute, *World Resources 1992-93: A Report by the World Resources Institute* (New York: Oxford University Press, 1992).

Asociacion Para el Avance de la Sociales en Guatemala, Growing Delemmas: Guatemala, the Environment, and the Global Economy, pamplet published by AVANCSO/PACCA, September 1992.

Pamela Falk, "U.S. Foreign Policy and the Latin America Time Bomb: Debt, Austerity, and the Limits of Stability," Journal of International Affairs, Summer/Fall 1989.

Rahna Rizzuto, International Planned Parenthood Federation, Western Hemisphere Region, Inc. pamplet, IPPF/WHR, 902 Broadway, 10th Floor, New York, NY 10010.

The World Bank, *The World Bank Anual Report 1993* (Washington D.C.: The World Bank, 1993).

The World Bank, *Anual Report, 1994*.

World Bank, *World Development Report 1990: Poverty* (Oxford: Oxford University Press, 1990).

Debt and International Finance Division of the World Bank's International Economics Department, *World Debt Tables* (Washington D.C.: The International Bank For Reconstruction and Development, 1991).

Conservation International, Conservation International Fact Sheet, Conservation International, 1015 18th street, NW, suite 1000, Washington, DC 20036.

CHAPTER 8

Burger (Mendes quote on page 13).

"Follow up: UN Conference on Environment and Development," UN Chronicle, September 1992.

Edward A. Parson, Peter m. Haas, and Mark A. Levy, "A Summary of the Major Documents Signed at the Earth Summit and the Global Forum," Environment, October 1992.

Statements on McDonald's paper cup.

Geoffrey Lean, *Atlas of the Environment* (Santa Barbara, CA: ABC-CLIO, Inc., 1994).

Joseph E. Daniel and Ilana Kotin, eds., *1993 Earth Journal: Environmental Almanac and Resource Directory* (Boulder, CO: Buzzworm Books, 1992).

"Green Parties on the Rise," USA Today Magazine, August 1990.

Telephone conversation with Portland City Hall clerk, Portland Oregon, January 12, 1994.

Seattle Solid Waste Utility, "Seattle's Road to Recovery: Seattle's Comprehensive Waste Management Strategies," information packet from the Seattle Solid Waste Utility, 710 2nd Avenue, Suite 505, Seattle Washington.

Telephone conversations with Austin City Hall offices, Austin TX, January 12, 1994.

Doreen Carvajal, Mother to a Rainforest, Austin American Statesman, January 17, 1995.

Aubrey Wallace, *Eco-Heros* (San Francisco: Mercury House, 1993).

Lean, Ghazi, and Lean.

Vice President Al Gore, *Earth in the Balance: Ecology and the Human Spirit* (New York: Penguin Group, 1993) (Mendes quote on page 286).